A Complete Guide to the Global Carbon Market

A Complete Guide to the Global Carbon Market

Profiting in a Low-Carbon World

Dr. M. A. Hashmi

Copyright © 2008 by M. A. Hashmi

Published by
MaxEnergy Inc., P.O. Box 201,
Mankato, Minnesota 56002, USA

All rights reserved. No part of this publication may be reproduced, stored in a retrieval system or transmitted, in any form, or by any means, electronic, mechanical, recorded, photocopied, or otherwise, without the prior permission of the copyright owner, except by a reviewer who may quote brief passages in a review.

Limit of Liability/Disclaimer of Warranty: The publisher and author have put best efforts in providing accurate and authoritative information in regard to the subject matter covered; they make no representation or warranties with respect to the accuracy or completeness of the contents of this book and specifically disclaim any implied warranties. No warranty may be created by the contents of this book. The advice and strategies contained may not be suitable for your situation. You should consult with a professional where appropriate. The publisher and author shall not be liable for any loss of profit or any other commercial damages, including but not limited to special, incidental, consequential, or other damages.

Printed in the United States of America

ISBN 978-0-578-01783-9

Contents

About the Author 9

Module 1: The Carbon Market 11

Module 2: Unique Features of Carbon Contracts 23

Module 3: Inspection and Verification of Carbon Credits Generation 29

Module 4: Global Climate Exchanges 37

Module 5: A Guide to the Green Financial Markets 45

Module 6: Global Carbon Industry Leaders 55

Module 7: Importance of Carbon Footprints 67

Module 8: Financial and Career Opportunities in the Carbon Market 73

Module 9: The Future of the Global Carbon Market 81

Module 10: Fifty Steps to Reduce Your Carbon Footprints 91

Appendix 1: Selected Acronyms 101

Selected References 107

List of Tables

Table 1.1:	Trends in U.S. Energy Data (Index 1990 = 100)	19
Table 1.2:	GHG Emissions Data for Polluting Nations Not Subject to Kyoto Reductions	20
Table 3.1:	Average Price of U.S. Carbon Credits Standards	31
Table 4.1:	Importance of OTC in Voluntary Carbon Market	44
Table 6.1:	World Bank Carbon Funds and Facilities	57
Table 6.2:	GHG Emissions by Sectors (IBRD and IDA Members)	59
Table 6.3:	Corporations Surveyed for the Sixth CDP Project	64
Table 9.1:	Total Emissions and Percent of Total U.S. Emissions from Regional Initiatives	90
Table 10.1:	Selected Websites on Reducing Carbon Footprints	98

I dedicate this book to my parents
who raised me in a low-carbon lifestyle and to my sons—
to whom the future belongs!

About the Author

DR. M. A. Hashmi is an international business consultant and professor of international business and finance at the Minnesota State University, Mankato. Dr. Hashmi has held several leadership positions and published over sixty applied research papers. Dr. Hashmi has contributed in developing a wind energy project and is currently studying wind energy development and greenhouse gas emissions reduction in Minnesota.

MODULE 1

The Carbon Market

THE global carbon credit market is expected to be one of the largest, if not the largest commodity market, by 2025. The market in GHG emissions could outstrip conventional commodities markets to become the most traded commodity, the head of the U.S. Commodities Futures Trading Commission said in June 2008 (Harvey 2008). The anticipated growth in the carbon market highlights the importance of proactive strategies to combat global environmental problems. A growing carbon market will create numerous economic opportunities for people and corporations. Some of these opportunities are highlighted in Modules 5, 8 and 10.

The carbon market provides a market-based mechanism for solving global environmental problems. This mechanism offers financial rewards to the corporations and countries taking steps to cut their greenhouse gas (GHG) emissions. At the same time, there is a cost to corporations and countries if they continue their business-as-usual policies regarding GHG emissions.

The six main greenhouse gases are:

- Carbon dioxide (CO_2)
- Methane (CH_4)
- Nitrous oxide (N_2O)
- Hydrofluorocarbons (HFCs)
- Perfluorocarbons (PFCs)
- Sulfur hexafluoride (SF_6)

Carbon credits are created by reduction of GHG emissions by a project compared with a predefined baseline. The tradable commodity of the carbon market is carbon credit. One carbon credit is equivalent to one metric ton of carbon dioxide equivalent (CO$_{2e}$) emissions of GHGs. There are different names for these carbon credits (depending on the nature and origin of the credits), and they do not enjoy uniform acceptance in all countries because of different certification standards, but carbon credits are a new currency of global climate programs (Victor and House 2004).

Carbon dioxide equivalent (CO$_{2e}$) is a quantity that describes the amount of carbon dioxide that would have the same global warming potential (GWP) of other five GHGs, when measured over a specified timescale. These six GHGs vary in their ability to trap heat in the atmosphere. Each GHG has a different global warming potential, which refers to its heat-trapping ability relative to that of carbon dioxide (CO$_2$). Corporations or countries mandated to reduce carbon emissions can purchase these carbon credits from other corporations or countries that have already satisfied their emissions targets and that are not under pressure to reduce their carbon emissions. Corporations or countries can also trade their carbon credits to satisfy voluntary targets in anticipation of future mandatory targets.

The price of carbon credits is determined by competitive bidding on one of the major climate exchanges, on over-the-counter (OTC) market, or by bilateral transactions. Carbon credit trading has real economic benefits to both developed and developing countries. The cost for developed countries to reduce each metric ton of CO$_{2e}$ in existing projects is over one hundred dollars, while developing countries' cost of creating carbon credits is ten to twenty-five dollars. Because of this cost difference, it is obviously very lucrative for developed countries to purchase carbon credits from developing countries rather than solely to reduce their own GHG emissions. Corporations from developed countries can purchase verified carbon credits from developing countries within a price range of five to twenty-five dollars and comply with the GHG emissions reductions goals. There are, however, limits on how many carbon credits from developing world can be traded. Carbon credits also offer significant benefits to the developing countries' low-carbon emitting projects by providing an additional source of revenue and boosting the economic feasibility of new green projects (WRI 2008).

Corporations and governments around the globe are projected to invest sixteen trillion dollars in new energy projects by 2030. Unless new renewable energy technologies are developed—in other words, if the status quo is maintained—most of the future energy projects may use fossil fuel, thus increasing GHG by another 50 percent(IEA 2008). Increased use of fossil fuel energy projects may create demand for offsetting low-carbon projects around the world. Most of the world governments have realized the intensity of the problem and are trying to strike a balance between energy needs and GHG emissions. The growth of the carbon credits market can be attributed to the desire of national governments to cut down GHG emissions and improve quality of life for their citizens.

According to the Intergovernmental Panel on Climate Change (IPCC), if GHG emissions are left unchecked, atmospheric warming will increase within a range of 2°F–11.5 °F (most likely range is 3.2°F–7.1 °F), by the end of the twenty-first century (IPCC 2007). These climate changes may result in weather-related catastrophes that potentially could disrupt the global economic system. An increase of 3.5 °F atmospheric temperature above the preindustrial levelcan cause irreversible damage to global ecological system. Among global environmentalists, there is a general consensus that governments must unite to combat the destruction of the global ecological system, and carbon trade is one of the mechanisms available to achieve this goal. Attempts to restructure global industries for lower GHG emissions and introduction of renewable energy technologiesmay be the most significant change after the industrial revolution.

It seems that the carbon credits trading mechanism is a reversal of the decade long policies of most governments, which had been subsidizing fossil fuel production. Fossil fuels (coal, oil, and natural gas) are the prime sources of GHGs. Governments across the world have been subsidizing the fossil fuel industry by an estimated amount of two hundred thirty-five billion dollars per year (Simms et al., 2004). A possible market solution to GHG emissions is the reduction or elimination of billions of dollars worth of governmental subsidies on fossil fuel production industries, thus increasing the cost of fossil fuels. The reduction or elimination of fossil fuel subsidies may potentially bring the cost of renewable sources of energy at par with traditional fossil fuel energy. In this scenario, governments and corporations will take proactive steps and voluntarily reduce their GHG emissions. This is not only an environmentally responsible strategy; it may also turn out to be a wise business decision by governments of the world, saving them badly needed trillions of dollars.

KYOTO PROTOCOL AND THE GLOBAL CARBON MARKET

The growth of the global carbon market is driven in part by initiatives like the Kyoto Protocol. The Kyoto Protocol to the United Nations Framework

Convention on Climate Change (UNFCCC) is an amendment to the original UNFCCC document that was agreed upon during the Earth Summit held in Rio de Janeiro in 1992. The purpose of the original UNFCCC document and the subsequent amendment (Kyoto Protocol) is to combat global warming by reducing the emissions of carbon dioxide (CO_2) and the other five greenhouse gasses. The Kyoto Protocol opened for signature on December 11, 1997, and has been ineffect from February 16, 2005. At the end 2008, there were 181 signatories of the Kyoto Protocol, representing about two-thirds of all GHG emissions in the world. Under the Bush administration, the United States (the largest GHG emitter in the world), decided not to ratify the treaty because the treaty did not mandate that China (the second largest GHG emitter in the world), Russia, Japan, and other developing countries cut down their GHG emissions.

Article 6 of the Kyoto Protocol describes the Joint Implementation (JI) projects, which are projects between developed countries. Carbon credits resulting from JI projects are called Emission Reduction Units (ERUs). Article 12 of the Kyoto Protocol legalizes the Clean Development Mechanism (CDM), which includes projects between developed and developing countries. Carbon credits resulting from CDM projects are called Certified Emissions Reductions (CERs). All these credits must be certified by an accrediting agency authorized by the UNFCCC, and the first ever accrediting agency—Norway-based Det Norske Veritas (DNV)—was authorized by the UNFCCC in March 2004. CDM and JI accreditation still lacks uniformity in standards and verification processes (WRI 2008).

The underlying principle of the Kyoto Protocol is "common but differentiated responsibility." All participating countries should pledge to reduce GHG emissions but the reduction targets may vary from country-to-country or region-to-region. For example, the EU has assumed heavier burden than the former Eastern bloc countries.

The Kyoto Protocol may not have been fully successful in reducing global GHG emissions because of nonparticipation of several emerging countries and the United States.For example, China has added almost as much GHG as the reduction made by the EU member countries in recent years. However, the Kyoto Protocol has had great success in constructing a fabulous framework of a global mechanism for lessening the impact of GHG emissions worldwide. This framework is capable of solving the present and future global environmental problems.

Before trading carbon credits, the seller must establish a baseline level and then estimate the reduction of CO_{2e} with reference to the established baseline. The reduction has to be verified by an accredited agency functioning under the guidelines of the UNFCCC or by other verification agencies in the voluntary markets. The host country's government must approve the selling of credits by the corporations or other entities operating on their soil. The baseline may significantly increase or decrease the quantity of carbon credits. Baselines can be project-specific, industry-specific, or country-specific. Getting experts to agree on estimates of carbon credits generation from a proposed project is often controversial and difficult; however, an accurate and widely accepted estimation of carbon credits quantity is vital for efficient pricing of underlying carbon credits. There is no widely accepted standard for estimation of baseline in voluntary markets. The compliance markets such as the Clean Development Mechanism (CDM) and EU Emission Trading Scheme (EU ETS) have better guidelines for estimating baseline. Broad categories of the investment projects that qualify for GHG emissions reduction include:

- Renewable energy and efficient energy supply projects
- Reduction of energy demand process or equipment
- Transport systems
- Waste management
- Forestry

World governments have been periodically meeting to resolve their differences relating to procedures and strategies for combating global environmental problems. These conferences are open to signatories and non-signatories of the Kyoto Protocol. Some of the key events are listed below.

Marrakech Accord

After two weeks of intense negotiations, the participating governments' representatives concluded the Marrakech Accord on November 10, 2001. The accord was a breakthrough, and it resolved concerns that countries from economiesintransition (particularly Russia) had raised about their GHG emissions reduction targets. The accord also ratified the rules and modalities of the Kyoto Protocol that dealt with the CDM and JI projects. The accord paved way for emissions trading as well.

Bali Action Plan

One hundred eighty countries adopted the Bali Action Plan after dramatic conclusion of the global environmental conference in Bali on December 15, 2007. The major achievement was the willingness of the U.S. delegate to look into numerical targets of GHG emissions reduction. Furthermore, the delegates discussed differentiated responsibilities for developed and developing countries. China and other emerging economies did inch forward, agreeing for the first time to seek ways to make "measurable, reportable and verifiable" emissions cuts. The Bali Action Plan has set a tone for the next round of global climate negotiation (post 2012). One of the most important agreements of the Bali Action Plan is as follows (UNFCCC 2008):

- Measurable, reportable, and verifiable nationally appropriate mitigation commitmentsor actionsby all developed countries
- Nationally appropriate mitigation actions by developing countries

Poznan Conference

Poznan conference ended on December 12, 2008. The conference (attended by representatives from one hundred ninety countries) made a little progress in reaching a consensus on key features of the second phase (post 2012) of the Kyoto Protocol, but the conference failed to achieve any breakthrough. The lack of progress can be explained by the global financial meltdown in October 2008 and the ending tenure of the United States' administration.

Copenhagen Conference

The Copenhagen conference is scheduled for December 2009. The sole purpose of the conference is to reach a general agreement on the second phase (post 2012) of the Kyoto Protocol. Delegates from participating countries hope Australia, Japan, and the United States may agree to participate in global climate accord at the Copenhagen conference.

THE UNITED STATES AND THE KYOTO PROTOCOL

While environmental groups have criticized the United States for its refusal to sign the Kyoto Protocol, China's GHG emissions have increased dramatically, a development that appears to support U.S. concerns. Meanwhile, the United States has reduced its GHG emissions relative to the gross domestic product (GDP). However, the reduction is not significant enough. The United States' climate change response will be much different under the Obama Administration.

The United States, under the Bush Administration, has criticized the Kyoto Protocol as unfair because it does not impose any GHG emissions reduction quotas on China and other polluting countries. The United States, however, has pledged to reduce its "carbon intensity" by 18 percent before 2012. Carbon intensity is the ratio of carbon emissions to economic activity, so it is a measure of economic efficiency and not a measure of actual CO_{2e} reduction. That is why the United States may still have a higher CO_{2e} emissions level by 2012, despite a reduction of carbon intensity (Eilperin 2005). The Bush administration's position was that the Kyoto Protocol will slow down economic growth and may cause job loss in the United States. Opponents of the Kyoto Protocol assert that the cost of implementing this agreement outweigh the benefits because mandatory reduction of GHG emissions will increase the cost of production in many industries and eventually hamper economic growth. The Bush administration estimated that the Kyoto Protocol's mandates would cost four hundred billion dollars to the United States' economy during the first phase (2008–2012) (*The Wall Street Journal* 2005). At the same time, the United States has agreed to join the open-ended dialogue after explicitly stating not to agree on Kyoto Protocol's GHG emissions reduction limits (*Dawn* 2005).

The United States has signed the Asia-Pacific Partnership on Clean Development and Climate, a pact that allows partnering countries to set their goals for reducing GHG emissions individually but with no enforcement mechanism. The Asia-Pacific Partnership on Clean Development and Climate, also known as AP6, is an international non-treaty agreement among Australia, Canada, India, Japan, the People's Republic of China, South Korea, and the United States. The Asia-Pacific Partnership on Clean Development and Climate was announced at an Association of South East Asian Nations (ASEAN) regional forum, and launched on January 12, 2006, at the partnership's inaugural ministerial meeting in Sydney (BBC News 2005). The United States is a signatory to the Sydney Declaration to the Climate Change, Energy Security, and Clean Development, to which Asia Pacific Economic Forum (APEC) members on September 8, 2007. The Sydney Declaration has stated a goal of decreasing energy intensity by 2030 from the 2005 level and increasing forest cover in APEC member countries by twenty million hectares by 2020 (SMH 2007). The Sydney Declaration fell short of mandating reduction of GHG emissions, but it was a step in the right direction.

The United States' position vis-à-vis GHG emissions reduction was to cut down GHG emissions without agreeing to mandates or firm targets. The United States' position can be better understood by analyzing key economic statistics and GHG emissions data (using 1990 as a base year) presented in Table 1.1; the year 1990 was also selected as a baseline year for the Kyoto Protocol emissions reductions. During 1990 through 2006, GHG emissions in the United States increased at an average annual rate of 0.9 percent, with a total increase of 15 percent. During the same period, GDP grew at an annual rate of 3.0 percent, electricity consumption grew at an annual rate of 1.9 percent, and population grew at an annual rate of 1.1 percent. On a positive side, the GHG emissions rate of growth is slightly slower than that for total energy or fossil fuel consumption and much slower than that for electricity consumption, population increase, or overall GDP growth. The data presented in Table 1.1 suggests that the U.S. corporations and public sector are aware of the environmental problems and are employing emissions reduction methodologies without any mandatory targets. Also, the slow growth of the GHG emissions can be attributed to the growth of the voluntary carbon credits market in the United States.

Table 1.1

Trends in U.S. Energy Data (Index 1990 = 100)

Year / Key Variables	1990	1995	2000	2001	2002	2003	2004	2005	2006	Average Annual Growth Rate
GDP	100	113	138	139	141	145	150	155	159	3.0%
Population	100	106	114	113	114	116	117	118	119	1.1%
Electricity Consumption*	100	112	127	125	128	129	131	134	135	1.9%
Fossil Fuel Consumption*	100	107	117	115	116	116	119	119	117	1.0%
Energy Consumption*	100	108	116	112	115	115	118	118	117	1.0%
GHG Emissions**	100	106	114	113	114	114	115	116	115	0.9%

*Energy constant weighted value.
** Global Warming Potential (GWP)–Weighted values

Source: EPA 2008

The Bush Administration's position vis-à-vis Kyoto Protocol was that the treaty is ineffective because several major countries were allowed to increase their GHG emissions while other countries were asked to cut down their emissions. Table 1.2 summarizes the change in GHG emissions data for the major polluting countries that are not subject to the Kyoto mandatory cuts. The time period selected is 1990 through 2004. The United States' GHG emissions increased by 16 percent while the Russian GHG emissions decreased by 32 percent. The alarming statistics are coming out of China and India. China, the second largest GHG emitter after the United States, has not taken appropriate measures to cut down their emissions, and emissions increased by 47 percent during the same period. Also, India's GHG emissions increased by 55 percent, and there are no sign of a reverse trend.

In 2004, the total Chinese GHG emissions were about 54 percent of the total U.S. GHG emissions. China has been building an average of one coal-fired power plant every week, and its governmentplans to continue doing so for years (Brahic 2007). Various predictions see China overtaking the United States in total GHG emissions sometimes before 2010. The Chinese government insists that the GHG emissions level of any given country is a multiplication of its per capita emission and its population. The Chinese and Indian GHG emissions statistics somewhat validate the United States position not to join the Kyoto Protocol, which has not targeted on the emerging economies' GHG emissions.

Table 1.2

GHG Emissions Data for Polluting Nations Not Subject to Kyoto Reductions

Country	GHG Emissions Change (1990–2004)
United States	+ 16 %
Russia	- 32 %
China	+ 47 %
India	+ 55 %

Source: Energy Information Administration 2008

The United States did adopt a conciliatory position at the Bali conference in December 2007 and agreed to numerical targets. In the 2008 presidential election, the candidates from both parties aggressivelysupported carbon reduction policies and proposed a carbon cap-and-trade system in the United States. A cap-and-trade system limits or "caps" total GHG emissions for a country (consequently toits industries). If a corporation reducesits GHG emissions below the cap, the country can sell the excess emissions permit to the corporation not able to reduce its GHG emission below the cap. Carbon trade creates cost and opportunities for domestic corporations.

President Barack Obama made an election campaign pledge to implement a cap-and-trade system to reduce GHG emissions. The ultimate goal is 80 percent reduction of GHG emissions below 1990 level by 2050. A key part of his program will be to require all pollution credits to be auctioned to ensure that U.S. industries pay for every ton of GHG emissions they release. The plan does address reengaging the United Nations to develop a global program. President Obama's plan is very comprehensive, addressing issues ranging from GHG emissions to increasing appliance efficiencies and introducing electric cars. Another part of this plan is to strategically invest one hundred fifty billion dollars to accelerate the commercialization of plug-in hybrids andthe development of renewable energy, to encourage energy efficiency, and to invest in clean coal plants and various other initiatives. The plan also ensures that 10 percent of the United States' electricity will come from renewable sources by 2012 and 25 percent by 2025. Legislators discussed a number of initiatives in 2008, and they are slated to continue the discussion in 2009. A recent study by the United States Environmental Protection Agency (EPA) concluded that a balanced carbon cap-and-trade program (details are found in Module 8) will not hurt the United States economy as feared by many policy makers (Power 2008). It is, ultimately, the decision of the new United States president and legislatures in 2009 to redefine the home and global GHG emissions market (Murray 2008). It is widely believed that the United States will become a major player in the global carbon market.

THE EUROPEAN UNION AND THE KYOTO PROTOCOL

The European Union (EU) has been a leader in the global carbon market. The EU implemented a cap-and-trade system and has been a driving force behind the Kyoto Protocol. EU has accepted a much higher emissions targets for itself without much cooperation from emerging economies and the United States. The EU has been able to cut down its GHG emissions while the United States, China, and India have continued polluting the environment. Between 1990 and 2004, the original EU—a group of fifteen countries—reduced its emissions by 0.8 percent. The GHG emission reductions are quite impressive among the countries that have newly joined the EU. The EU group of 23 Nations (EU-23) has reduced its emissions by 5 percent during the same period.

The EU has been an advocate of tougher measures to reduce GHGs by agreeing to reduce its emissions by 8 percent by 2012 using a baseline of 1990 emissions levels from the original fifteen EU countries. This is a greater reduction target than the global average of 5.1 percent GHG emissions reduction. However, the newly admitted twelve countries raised the baseline, so the EU does not have to cut GHG emissions as much as it originally estimated (Eilperin 2005). EU officials have created industry-wide quotas for GHG reduction, targeting the six key industries of steel, energy, cement, glass, brick making, and paper. The EU imposed a punitive tax up to forty euros per ton in 2005 and one hundred euros per ton of CO_{2e} in 2008 if a member nation failed to meet their industry-wide quotas.

In early 2005, the European Union implemented a cap-and-trade system before the Kyoto Protocol called the Emission Trading Scheme (EU ETS). This program is designed to transfer governments' emissions reduction obligations to the private sector by forcing domestic corporations and multinational corporations to reduce their GHG emissions before the Kyoto Protocol reductions were set to be enforced in 2008 (Merril and Vivek 2005). These transactions are also called "allowance-based" transactions compared to "project-based" transactions as mandated by the Kyoto Protocols' emissions reduction guidelines. In this regard, Europe has taken a lead, and the EU ETS is a step in the right direction and well ahead of Kyoto Protocol enforcement (2008–2012). As of 2008, the EU is the leader in the carbon market, and the European corporations are the dominant players, trading carbon credits in the global climate exchanges and the over-the-counter markets.

Module 2

Unique Features of Carbon Contracts

THE carbon credits market is highly fragmented because various greenhouse gas (GHG) emissions reduction programs apply different standards to estimate and verify carbon dioxide equivalent (CO_{2e}) reductions. Not all carbon credits can be interchanged or offset by carbon credits generated by other programs and countries. Each type of carbon credits represents a unique legislative environment and geographical location. There is a need to improve linkage and interchangeability of various carbon credits and, eventually, to create a global carbon market. The unique nature of most of the carbon credits is explained in the following section.

1. **Certified Emissions Reductions (CERs)**

    Certified Emissions Reductions (CERs) are the carbon credits generated by a Clean Development Mechanism (CDM) project—a carbon abatement project in a developing country following the requirements of the Kyoto Protocol as defined in Article 12. CDM projects cover a wide range of GHG emissions reduction projects such as

    - Fuel switching
    - Energy efficiency
    - Transportation
    - Forestation
    - Renewable energy (wind, solar, geothermal, hydro, biogas, and biofuels)

    These carbon credits are validated and registered through the CDM Executive Board (EB). After registration, these projects can be traded at a climate exchange or at the over-the-counter (OTC) commodity market. At the end of 2008, China was the largest producer of CERs and accounted for about half of all CERs generated, followed by India, Brazil, South Korea, and Mexico. Purchasers of these credits were mostly European corporations and governments. Most of the Chinese projects were large in size, whereas India specialized mainly in small CDM projects.

Most of the CDM projects would not be feasible without the revenue generated by CERs sale. It is worth mentioning that all intended CDM projects may not eventually be classified by the CDM Executive Board as CDM projects. In some instances, project design methodology can be rejected by the country's Designated Operational Entity (DOE) if the proposed methodology does not reduce carbon emissions or the project does not pass the "additionality" requirement. The "additionality" requirement is enforced to make sure the carbon reduction projects are developed in addition to (and not in lieu of) projects employing existing high CO_{2e} emissions methodologies. These failings can result in the project not being approved by the CDM Executive Board. Finally, the CDM projects may not be classified as CDM projects if the project encounters undue delays in the implementation phase; in this instance, the emission credits cannot be certified and do not become a CER. The uncertified credits, however, can be traded as Verified Emission Reductions (VERs), as explained later in this section.

2. **Emission Reduction Units (ERUs)**

   Emission Reduction Units (ERUs) are the carbon emissions reduction under the Joint Implementation (JI) projects. The JI projects are carbon abatement projects in economies of transition, mostly former Soviet Bloc countries. JI projects operate in accordance to Article 6 of the Kyoto Protocol, and these carbon credits are traded between two developed countries. As of 2008, ERUs represent a small fraction of the total global carbon credits. The JI project approval and emission certification process is somewhat similar to the CDM. ERUs are not traded on global climate exchanges.

3. **Assigned Amount Units (AAUs)**

   All developed countries (Annex B countries) that are signatories of the Kyoto Protocol are allocated maximum GHG emissions quantity permits for the period of 2008 to 2012. The countries are allowed to trade these credits among themselves based on their need and supply of these units. These tradable carbon credits are called the Assigned Amount Units (AAUs), which are traded at the intergovernmental level.

4. **European Union Allowances (EUAs)**
European Union Allowances (EUAs) were created as a result of the EU Emission Trading Scheme (EU ETS). EU ETS functions under the EU cap-and-trade mechanism, which has existed since 2005, well before the Kyoto mechanism was introduced. Member countries and their designated industries are allowed to trade their carbon allowances to other member countries and industries.

The first compliance phase was 2005 to 2007. The second compliance phase coincides with the Kyoto mechanism, in other words, from 2008 to 2012, while the third compliance phase covers the 2012 to 2020 period. In the second phase (2008–2012), there are limits on how many CERs and ERUs can be traded to comply with the EU ETS cap. These trading restrictions put more of a burden on the EU countries to reduce emissions themselves or to trade EU ETS among themselves, without the benefit of purchasing cheaper CERs. At present, EU ETS phase III (2012–2020) stipulates more stringent requirements compared to phase II requirements; however, post-Kyoto (beyond 2012) developments will redefine the requirements of EU ETS phase III.

The EU ETS mechanism covers all twenty-seven member countries and includes all major polluting industries—for example, power and heat, refineries, metal, mineral, oil and gas, cement, lime and glass, and pulp and paper. The total cap for all twenty-seven countries is 2.1 gigaton (billion ton) carbon dioxide equivalent per year ($GtCO_{2e}$/year), between 2008 and 2012. The emission allowance is distributed to member countries, and within the countries, the allowance is distributed to various sectors.

The EU ETS mechanism was linked to the CDM market beginning in 2005 and linked with the JI market in 2008. This suggests that, within a specified limit, EUAs can be traded with CERs and ERUs. There are also links to other non-EU member countries, such as Norway, Japan, and the United States, so EUA can be traded outside the EU member countries. A wide range of EU ETS contracts such as spot, forward, futures, options, and swaps are traded on climate exchanges, by brokers on OTC markets and bilaterally among the interested parties. Most of the EUAs are traded on the London-based European Climate Exchange (ECX) and at the OTC market. EU ETS represents the largest group of carbon credits traded on the ECX and OTC markets and is expected to comprise around 70 percent of all classes of carbon products.

The price of the EUAs is determined by the allocated quota of each country and the annual certified emission report of each country. If countries are not short of their emission quota, the price will decline. EU ETS price also influences the CDM secondary market pricing because EU countries are the major buyers of CERs generated by CDM projects.

5. **Verified Emission Reductions (VERs)**

    Verified Emission Reductions (VERs) are the carbon credits for the voluntary markets in the United States and around the world. These markets are functioning outside the realm of the Kyoto Protocol; thus, the VER credits are verified by a mutually agreed upon third party and not by the UNFCCC. VER credits are usually generated by small projects. Some of the VERs are the credits that may have failed to be classified as CERs because the CDM projects could not meet "additionality" and "leakage" requirements. Traders are involved in the VER market either because of their investment and corporate philosophy (voluntary market) or because they are anticipating new emission compliance caps in the near future.

    The Voluntary Carbon Standard (VCS) is a global benchmark standard for measurement and recognition of VERs. VERs can be traded on the OTC market by interested parties to meet their voluntary targets. Buyers usually pay a discounted price for VERs compared to CERs because VERs are approved by somewhat relaxed verification and monitoring procedures compared with the CERs.

6. **Emission Reductions (ERs)**

    Emission Reductions (ERs) are carbon credits that are not validated or certified by a Designated Operational Entity (DOE), but these credits can be mutually traded. The market for ERs is much smaller than the market for CERs and other certified carbon products. ERs are quite often traded on a voluntary market across the world.

7. **Removal Units (RMUs)**

   Removal units are related to land use, land use change, and forestry (LULUCF) activities, and one unit is equivalent to one metric ton of CO_{2e}. RMUs can be converted to the Assigned Amount Units (AAU) within the national registry and cannot be banked for a future commitment period. RMUs represent a very small portion of the carbon market, and these credits can be created from projects around the world.

8. **Carbon Financial Instruments (CFIs)**

   The overall nature of the U.S. carbon market is voluntary. There are noticeable voluntary and pre-compliance activities in the U.S. voluntary carbon market. The Chicago Climate Exchange (CCX) has been trading carbon products since 2005. The CCX tradable commodity is the Carbon Financial Instrument (CFI), which consists of one hundred metric tons of CO_{2e}. The voluntary nature of the U.S. market is soon to be changed because of several regional climate control initiatives and the highly anticipated federal cap-and-trade system in the United States.

9. **Voluntary Carbon Units (VCUs)**

   The Voluntary Carbon Standard (VCS) provides robust standards for the global voluntary carbon market. A second version of the VCS was introduced in November 2007. The VCS has created a new voluntary market currency called the Voluntary Carbon Unit (VCU), with the sole purpose of providing a degree of standardization to the voluntary carbon market by achieving "real, measurable, permanent, additional, independently verified, and not double-counted" emissions reductions (VCS 2009). The VCS was developed by the International Emissions Trading Association (IETA), The Climate Group (TCG), the World Business Council for Sustainable Development (WBCSD), and the World Economic Forum (WEF). VCUs are for global voluntary market projects and show the project administrators' voluntary market willingness to standardize their practices and, hence, the price of voluntary market's carbon credits.

Module 3

Inspection and Verification of Carbon Credits Generation

THERE is a huge need for professionals who can understand the carbon emissions reduction processes, develop carbon emissions reduction strategies, and verify the accuracy and sustainability of carbon credits generation in a project. Verification of carbon credits generation and development of carbon strategies are complicated tasks because the standards and regulations vary among all major carbon markets; therefore, different expertise is needed in the EU ETS, CDM, JI, and other voluntary markets. Some of the most stringent standards of verification are enforced in the CDM projects under the Kyoto Protocol (2008–2012). In each voluntary market, participating corporations or industries decide the scope and verification criteria. In the United States' voluntary market, the process of verification may become more cumbersome as a result of the three regional GHG reduction initiatives, along with the anticipated federal cap-and-trade program.

The carbon credits verification process requires complete transparency from project developers. All documents and data must be verifiable. Estimates of carbon credits generation should be accurate and comparable to other projects in the same industry. Each project should also provide a detailed account of all identifiable emission sources within the energy, industrial process, and waste sectors. Finally, the carbon emissions estimates should be consistent over time.

Carbon credits are a marketable commodity, and for most projects, the revenue generated from carbon credits is crucial for a project's success. In order to avoid rejection by the CDM Executive Board or other authenticating agencies, project developers should make an effort to offer the necessary data for verification.Project developers must provide complete and accurate information, and project methodologies must be in accordance with the guidelines of the carbon credits approval authority. Project developers must set up sound monitoring plans and appropriate quality control procedures. Above all, they must ensure that all activities are well documented according to established industry standards.

BENEFITS OF VERIFICATION PROCESS IN THE U.S. MARKET

The verification standards for GHG emissions reduction vary from country to country depending on the nature of the carbon reduction mechanism. As of early 2009, there are no uniform verification standards in the U.S. voluntary markets. Carbon credits generated from the markets with well-defined and verifiable GHG emissions reduction standards usually price well by the investors.

One of the most prestigious verification standards in the voluntary market is the Gold Standard, which incorporates sustainability variables, along with other GHG emissions reduction standards. Project developers must strive to obtain the seal of approval from the Gold Standard. The Gold Standard can significantly increase the price of the carbon credits generated from the project, thus bringing increased revenue stream. The Gold Standard Foundation is a nonprofit organization based in Switzerland that is supported by about sixty global nongovernmental organizations (NGOs). The Gold Standard Foundation formulated the Gold Standard approval methodology for CDM projects in 2003 and for voluntary offset projects in May 2006. The Gold Standard seal gives confidence to carbon credits investors and host countries that an underlying project is a socially responsible and sustainable energy project.

At the end of 2008, there were over fifty projects in the process of obtaining the Gold Standard's seal of approval, and another dozen projects had already been approved. The Gold Standard has emerged as a global brand that offers a premium on carbon credits bearing its seal of approval. Most of the Gold Standard projects focus on renewable energy, energy efficiency, and sustainable impact on communities. Furthermore, projects eligible for the Gold Standard seal of approval should provide evidence that the project is not a routine project (additionality requirement) and that the project's primary intent is to reduce GHG emissions (Gold Standard 2008). Table 3.1 has listed average price of carbon credits for a selected few standards in the U.S. carbon market.

Table 3.1

Average Price of U.S. Carbon Credits Standards

Carbon Credits Standard	Price (July–Aug. 2008)
Gold Standard	$ 15.83
California Climate Action Registry	$ 10.76
Voluntary Carbon Standard	$ 7.25
Other credits	$ 4.24
Chicago Climate Exchange	$ 3.90

Source: Ball 2008b

The U.S. carbon credits market is not uniformly regulated; that is why the price of various standards deviates significantly. The most revenue generating carbon credits are the one with the Gold Standard's seal of approval. Green project developers are encouraged to follow Gold Standard's criteria, if possible. The California Climate Action Registry (CCAR) also priced well in the third quarter of 2008. TheCCAR is a voluntary program to help corporations and organizations in California to register and inventory their GHG emissions. The price of CCAR credits is higher because the registry provides a verified set of GHG emissions data from reporting entities, along with well-defined accounting and verification standards.

The newly created Voluntary Carbon Standard was priced at about half of the Gold Standard price (better regulated), and the price of the Voluntary Carbon Standard was almost double the least-regulated Chicago Climate Exchange carbon credits price. The carbon credits at the Chicago Climate Exchange were the lowest revenue generator because the investors have difficulty verifying the accurate quantity of GHG emissions reduction (Ball 2008b). It is obvious from the price data that carbon credits generated from better-regulated and verified projects are priced higher and, therefore, generate more revenue for the project (Table 3.1).

The Certification and Verification Process of CDM Projects

The certification process of the Clean Development Mechanism (CDM) projects is the most demanding and complicated. The objective of the process is to ensure that the carbon credits generated from projects in developing countries both satisfy criteria established by the Kyoto mechanism and significantly reduce GHG emissions. The Kyoto mechanism also set up a number of entities directly responsible for this certification process. The role of these entities and the certification process (CDM 2008) is explained in this section.

1. *Designated National Authority (DNA)*
    To participate in a CDM project, a developing host country must authorize a domestic entity to approve carbon credits generated from the project. The host country's Designated National Authority (DNA) issues a letter of approval (LoA) to transfer the host country's carbon credits to international investors or organizations. After approval from the developing country's DNA, the carbon credits generated from a CDM project must be approved by the investor's country as well.
2. *Designated Operational Entity (DOE)*
    A Designated Operational Entity (DOE) is an international or host country entity with the sole objective of validating carbon credits from a proposed CDM project. The DOE needs to be accredited by the CDM Executive Board, and it can operate in multiple countries after accreditation. The DOE not only validates and requests registration of the carbon credits generating project, but it also regularly verifies emission reductions after the project is operational.
    The project developer provides a project design document (PDD). The project design document is developed using the format specified by the CDM Executive Board. The project design document must include the baseline selection methodology, the monitoring procedure, the credit generation period, the environmental impact, and comments from all stakeholders of the project. The DOE validation process and the assessment by the host country's DNA can be carried out at the same time.
3. *CDM Executive Board*
    Carbon emissions reducing projects in the developing world are called CDM projects. After CDM projects are validated by the DOE of the

country, these projects are registered through the CDM Executive Board, which is a centralized global entity functioning under the auspices of the United Nations. After the CDM Executive Board registers the project, the project can start generating tradable Certified Emission Reductions (CERs). The CDM Executive Board has delegated the registration task to the secretariat of the United Nation Framework Convention on Climate Change (UNFCCC).

4. *Certification Procedure*

The above-mentioned entities play crucial roles in the certification process for CERs from a developing country's CDM project. The developer prepares an application for national approval of the project and submitsit to the country's Designated National Authority (DNA). The DNA ensures that the project complies with national sustainable development criteria and local environmental requirements. The DNA must verify that the CDM project activity is "additional"if GHG emissions are reducedbelow those that would have occurred in the absence of the registered CDM projectactivity.

The Project Design Document (PDD) is prepared and submitted to the DOE, a third party verification and validation corporation accredited by the UNFCCC. The PDD is developed using a specific format recommended by the CDM Executive Board, which includes a detailed description of the project, GHG emissions reductions, baseline methodology, crediting period, monitoring methodology, environmental impacts, and various stakeholders' statements.

Validation of a project is then carried out by the DOE. This could be done in parallel with the assessment by the host country's DNA. Once the CDM Executive Board completes the review and registers the project, the project can start operations and accumulate GHG reductions as CERs. After the project is operational, all activities and data documented during the monitoring stage are used for the subsequent verification and issuance of CERs. Data quality audits and performance reviews need to be conducted by a verification agency. The DOE verifies and certifies the emissions reductions on an annual or biennial basis. Once the GHG reductions are verified and certified, CERs are issued by the CDM Executive Board. The certification process is summarized in Figure 3.1.

Figure 3.1

Certification Processes of CDM Projects

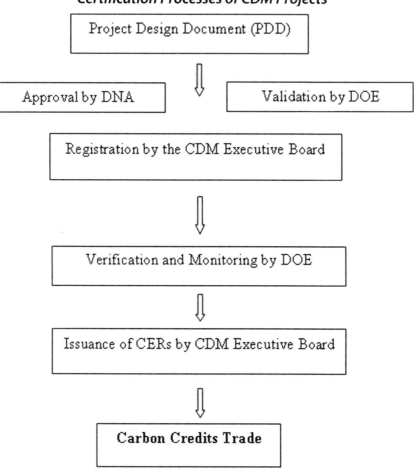

The Certification and Verification Process of JI and Voluntary Market Projects

The Joint Implementation (JI) projects are among the Annex B countries (the EU, Norway, Switzerland, and most of the former Eastern Bloc countries), and they are certified by a somewhat similar mechanism as the CDM projects. JI projects are also subject to the "additionality" requirement, which means that the project must be additional to what would otherwise have occurred. JI projects cover similar carbon reduction methodologies (energy efficiency, fuel switching, renewable energy, and forestation) as the CDM projects. A JI project's carbon credits (ERUs) are issued by the host country, and these credits are tradable as of 2008 (UNFCCC 2008).

There are two levels of certification for JI projects. The "track 1" procedure is for projects that have met all of the eligibility requirements. The "track 2" procedure is applied to projects that partially meet the eligibility requirements; thus, an independent entity accredited by the JI Supervisory Committee (JISC) has to determine whether the relevant requirements have been met.

Voluntary market carbon credits do not go through as rigorous a certification process as CERs and ERUs. Corporations or projects need to set up a baseline, and then they report their carbon abatement quantity on a regular basis. These credits are noncompliance credits and are traded among the corporations participating in voluntary markets. The voluntary markets' carbon credits are sold at large discounts compared to the CERs and ERUs. Corporations are generating carbon credits in the voluntary market for managing their carbon exposure for non-regulatory purposes. The carbon credits generated from the U.S. voluntary market are not frequently traded in international carbon trading platforms (exchanges and the OTC markets), which generally operate under the Kyoto Protocol's requirements of monitoring, verification, validation, and certification. The voluntary markets' carbon credits are called by different names in different markets. Some of the carbon products traded on voluntary markets are VERs, ERs, CFIs and other regional carbon credits.

Module 4

Global Climate Exchanges

GLOBAL climate exchanges offer an efficient carbon trading mechanism. These exchanges also offer a cost-effective and transparent method in which the rules are known to each party. Climate exchanges also offer an anonymous mechanism where buyer and seller do not know each other. Such an arrangement is particularly appealing to privacy-minded investors, such as hedge funds. Climate exchanges still account for about one-third of all carbon contracts. Some of these exchanges exclusively deal in carbon contracts, while other exchanges were traditionally commodity exchanges and lately have decided to offer carbon products along with other traditional commodities (Cundy 2008). There are over a dozen climate exchanges, and the market has witnessed consolidation and cooperation in the industry. By far, the European Climate Exchange (ECX) is the largest climate exchange in the world (ECX 2008).

As of 2008, three types of carbon credits have accounted for most of the trade on the exchanges: European Union Allowances (EUAs), Certified Emission Reductions (CERs), and generally noncertified voluntary credits or Voluntary Emission Reductions (VERs). The EUAs represent the largest segment on the climate exchanges. A brief description of some of the major climate exchanges follows:

1. **The Chicago Climate Exchange (CCX):**
   (www.chicagoclimatex.com)
   The Chicago Climate Exchange (CCX) is the first ever—and North America's only—voluntary, legally binding climate exchange. The CCX primarily offers carbon transactions for projects in North America and offset projects in Brazil. The exchange has adopted an independent verification mechanism for all six greenhouse gases (GHGs) and has been in

business since 2003. As of June 2008, there were about four hundred full members, participating members, and associate members at the CCX ranging from multinational corporations, educational institutions, farm groups, cities, and state governments. Among the twenty-eight founding members are Ford, DuPont, BP America, American Electric Power, and other leading entities (CCX 2009). The founding members of the CCX emit more CO_{2e} than the United Kingdom. In the absence of mandatory cuts, setting up the CCX and announcing emissions reduction were great voluntary initiatives in the United States (Khanna, 2003). The CCX tradable commodity is the Carbon Financial Instrument (CFI), which consists of one hundred metric tons of CO_{2e}. CCX has developed standardized rules for all major industries, including renewable energy, methane emission from agriculture, coal mines, landfills, forestry, and ozone depleting substance destruction.

CCX membership has three membership categories: member, associate member, and participative member. Members were committed to reduce their CO_{2e} emissions by a percentage point each year during phase I (2003–2006) compared to their baseline level (average of their 1998–2001 carbon emissions). Phase II parameters extended the reduction period through 2010, with an additional 2 percent reduction commitment for Phase I members and a total of 6 percent reduction commitment by 2010 for new members joining in Phase II. Members get a credit for the carbon emission reduction at a future date. The National Association of Security Dealers (NASD) oversees the carbon credit clearing mechanism at the CCX. In addition to the GHG trading program, the CCX has established the Chicago Climate Futures Exchange (CCFE), a wholly owned subsidiary, which currently offers standardized and cleared carbon futures contracts and sulfur dioxide emission allowance contracts. The Carbon Financial Instrument (CFI) future contract's performance at the CCX is gauged by a newly created emissions index.

The Chicago Climate Exchange has taken the lead in helping to create more climate exchanges to satisfy other markets' needs (CCE 2009). At the beginning of 2009, CCX has granted affiliate status to four international climate exchanges:

- European Climate Exchange (ECX)

- Insurance Future Exchange (IFEX)
- Montréal Climate Exchange (MCeX)
- Tianjin Climate Exchange (TCX)

CCX management is planning to grab a large share of the U.S. market when the United States adopts a cap-and-trade system or any other Kyoto-like regime; thus, the CCX will compete with the New York Mercantile Exchange (NYMEX) and other carbon exchanges.

2. **The New York Mercantile Exchange (NYMEX):** (www.nymex.com)
   The New York Mercantile Exchange (NYMEX) is the world's largest commodity exchange, and its decision to launch carbon products starting on March 17, 2008, signaled the growth and importance of the carbon market. The first few contracts NYMEX offered were future and options contracts on EUAs and future contracts on CERs. VER contracts were added in the following months. NYMEX also offers some U.S.-specific carbon products such as U.S. sulfur-dioxide emission allowance and U.S. nitrogen-oxide emission allowance contracts (*Carbon Finance* 2008b). The CER and EUA contract size is one thousand tons of CO_{2e}. It is premature to analyze the size and performance of these carbon contracts at present because the carbon products were available only since March 2008.

   NYMEX is offering carbon products as part of the Green Exchange initiative, which is a partnership between NYMEX and Evolution Market, a New York-based brokerage house (Green Exchange 2008). The contracts will be listed at NYMEX until the Green Exchange is fully developed by 2009. In the end of 2008, NYMEX is offering U.S. Regional Greenhouse Gas Initiative (RGGI) carbon allowance futures and options contracts, Voluntary Carbon Units (VCUs) contracts, and Verified Emission Reductions (VERs) contracts. The Green Exchange has worked with environmental organizations and financial institutions to develop voluntary carbon market and renewable energy standards, which are essential for risk free transactions on the exchange. The Green Exchange is a 100 percent carbon neutral corporation.

3. **The European Climate Exchange (ECX):**
   (www.europeanclimateexchange.com)
   The European Climate Exchange (ECX) is the largest climate exchange

in the world and captures a large segment of the European carbon market. ECX is a member of the Climate Exchange Plc group of corporations, and trading on ECX began in January 2005. Climate Exchange Plc (CLE) is listed on the AIM market of the London Stock Exchange. Other member corporations include the Chicago Climate Exchange (CCX) and the Chicago Climate Futures Exchange (CCFE).ECX has also established a strategic alliance with the International Petroleum Exchange in London.

As of early 2008, the ECX traded over one billion metric tons of CO_{2e}. More than ninety major businesses such as ABN AMRO, Barkleys, BP, Shell, and Fortis have traded their carbon products on the ECX. Key carbon products on ECX are EUA futures and options and CER futures and options. The exchange also clears the over-the-counter (OTC) market's futuresand options contracts (ECX 2008).

Many ECX contracts take place in a regulated electronic marketplace provided by theICE Futures. ICE Futures is a whollyowned subsidiary of Intercontinental Exchange, Inc., which operates the global OTC marketplace for commodity trading on its Internet-based interchange trading platform. ICE Futures trades standard contracts with clearing guarantees provided by LCH Clearnet; trade started on April 22, 2005.

4. **The Canadian Climate Exchange (CCE):**
(www.canadianclimateexchange.com)
Canadian Climate Exchange Inc. (CCE) is a wholly owned subsidiary of ICE Futures Canada (formerly known as Winnipeg Commodity Exchange). ICE is the fully electronic commodity derivative exchange in North America. CCE was created in 2003 to trade carbon products in Canada. The carbon trading never materialized at CCE because of a declining number of carbon credits and no clear policy of the Canadian government. One of the confusions was created by the government's guarantee to purchase carbon credits at C$15 per ton irrespective of the global price. A higher global price will obviously create an arbitrage opportunity for Canadian contracts. Canadian authorities are unwilling to let Canadian contracts to be traded on global climate exchanges (CCE 2008).

5. **The Montréal Climate Exchange (MCeX):** (http://www.m-x.ca)
On December 7, 2005, the Chicago Climate Exchange and the Montréal

Exchange, the oldest exchange in Canada, agreed to a Canadian environmental product market called the Montréal Climate Exchange (MCeX) targeting Canadian carbon credits contracts (Environmental Finance 2005). The exchange is designed to develop trading, clearing, and registry services for the Canadian market. Carbon contracts are expected to be traded after going through Canadian regulatory approval. The exchange will offer large Canadian industrial corporations and power plants a platform to trade carbon credits among themselves. The Montréal Climate Exchange will explore other environmental contracts (water and weather) but will initially concentrate on carbon products (MCeX 2007).

6. **Nord Pool:** (www.nordpool.com)

   Nord Pool is an Oslo-based climate exchange offering EUA spot and forward contracts, along with CER forward contracts. Nord Pool was the first ever exchange to list EUAs (February 2005) as well as CERs (June 2007). The OTC-traded EUA and CER contracts can also be cleared by Nord Pool Clearing, the contract clearing platform. It is still a small exchange with total volume of 120 metric tons of CO_{2e}. At the end of 2007, there were one hundred twenty-five members from eighteen countries. In early 2008, Nord Pool was acquired by OMX Nordic Exchange (a Swedish exchange group). It is speculated that OMX might be merging with the United States' NASDAQ exchange, so Nord Pool may have a transatlantic clientele base in coming years.

7. **Bluenext:** (www.bluenext.eu)

   Bluenext is a joint venture of the New York Stock Exchange Euronext (60 percent) and the French Caisse des Dépôts (40 percent). The exchange was incorporated on December 21, 2007. The exchange offers EUA spot and CER spot contracts and is planning to offer EUA and CER futures. Bluenext is a Paris-based climate exchange, which was previously a part of Powernext, a French energy exchange; thus, it has a large support from power-related French corporations. Bluenext claims to have the most efficient trading platform, where money can be available within fifteen minutes after a deal closes. Another important feature is Bluenext's ability to identify the source of the CERs; this is valuable information because certain CERs are not fully accepted by all parties and may have a questionable certification process, country of origin, or type of CDM project.

8. **Climex:** (www.climex.com)

    Climex started in 2003 as a small Netherland-based climate exchange. Climex has several distribution partners in the Netherlands, Hungary, and Australia. In 2008, all six Climex Alliance partner exchanges merged to form Climex, a pan-European entity. Climex offers EUA and VER spot contracts. Climex also auctions carbon credits through its Climex Auction Platform. A wide range of carbon products are auctioned on this platform.

9. **The European Energy Exchange(EEX):** (www.eex.com)

    The European Energy Exchange (EEX) is a German exchange that partners closely with Frankfurt-based Eurex, one of the largest financial derivative exchanges in the world. EEX offers EUA spot and futures contracts, CER futures contracts, and options on EUA futures contracts. The joint membership base of EEX-Eurex is about six hundred trading participants, and the exchange enjoys the benefits of fully integrated trading and clearing of exchange-traded contracts and OTC transactions based on known financial market standards. In addition, EEX-Eurex clears bilaterally agreed OTC transactions just like a regular exchange transactions.

10. **The Asia Carbon Exchange (ACX):** (www.asiacarbon.com)

    The Asia Carbon Exchange (ACX; operated by Asia Carbon Group) is based in Singapore, and it is the major Asian hub for the CERs generated by CDM projects in Asia. This exchange has organized several auctions for CERs and VERs. The exchange has physical presence in many countries, but it is still a small player in the global carbon trade.

11. **The Hong Kong Exchange (HKX):** (www.hkex.com.hk)

    The Hong Kong Exchange (HKX) is one of Asia's largest financial exchanges, and the exchange is planning to offer carbon products in cooperation with the Tianjin Climate Exchange (TCX) by 2009. HKX will offer CER trading and auctions, along with services for green corporations' IPO and financing needs. HKX has a potential to capture the Chinese CDM market because many green Chinese corporations have cross-listed their stocks on the HKX, and the Hong Kong government has entered into a carbon equivalency agreement with the neighboring Chinese province of Guangdong.

12. **Tianjin Climate Exchange (TCX):**
    (www.tianjinclimateexchange.com)
    The Tianjin Climate Exchange (TCX) is a joint venture of China National Petroleum Corporation Assets Management Co. Ltd. (CNPCAM), the Chicago Climate Exchange, and the City of Tianjin. The exchange plans to offer climate management solutions to Chinese corporations and to start trading CER contracts.
13. **Bolsa de Mercadorias & Futuros (BMF):** (www.bmf.com.br)
    Bolsa de Mercadorias & Futuros (BMF) is the largest Brazilian exchange based in São Paolo. This exchange regularly conducts CER auctions generated from Brazilian CDM projects. The Brazilian Carbon Market (MBRE) is a joint initiative by BMF and the Brazilian Ministry of Development, Industry, and Foreign Trade (MDIC), whose main objective is to develop an efficient trading system for environmental products aligned with the principles underlying the Kyoto Protocol.

    The first phase of the MBRE was the implementation of the BMF Carbon Facility, which hosts the registration of projects validated by the Brazilian Designated Operational Entities (DOE) according to the CDM criteria. The second phase of the MBRE focuses on the development and setup of a web-based electronic trading platform for carbon credits auctions. This system enables the trading of carbon credits generated by CDM projects. A specific forward market environment will also be implemented for trading carbon credits that are still in the process of generation and certification.

THE OTC MARKET AND BILATERAL EXCHANGE

The over-the-counter (OTC) market accounts for over half of all carbon contracts. This market is dominated by brokers who enjoy intimate knowledge of the carbon market. These brokers often verify authenticity and origin of carbon contracts and facilitate transactions between sellers and buyers, who may know each other. Many trading partners prefer the human contact despite the fact that transactions on the OTC market are more expensive than transactions on climate exchanges. Buyers and sellers of carbon products can bypass exchanges, and brokers may mutually trade various types of carbon products. The process is carried out on the broker's Website or by personal contacts. A small portion of contracts are negotiated bilaterally between seller and buyers.

The OTC and bilateral contracts can be cleared through a climate exchange if desired. The OTC market is offering new opportunities such as a "secondary trading" mechanism for the CERs. These new offerings have made this market competitive with the climate exchanges, but the traders prefer organized climate exchanges when trading the CERs.

The OTC market dominates the European EUAs trade, and traders on the global voluntary carbon market also favor the OTC market transactions. The volume in the global voluntary carbon OTC market nearly tripled in 2007 compared to 2006, to 42.1 million tons of carbon credits. Combined with the 22.9 million metric tons transacted on the CCX in 2007, a confirmed total volume of 65 million metric tons was transacted in the voluntary carbon market in 2007. The total value of the global voluntary carbon market was $330.8 million, and the OTC market share was $258.4 million (78 percent). There is no authentic data available for 2008 global voluntary carbon market, but initial estimates suggest a modest increase in the volume of the global voluntary market credits. Details of the global voluntary carbon market are summarized in Table 4.1.

Table 4.1

Importance of OTC in Voluntary Carbon Market

Markets	Volume* 2006	Volume* 2007	Value** 2006	Value** 2007
Voluntary OTC Market	14.3	42.1	58.5	258.4
Chicago Climate Exchange (CCX)	10.3	22.9	38.3	72.4
Total Voluntary Market	24.6	65.0	96.7	330.8

*Metric tons of CO_{2e}
**Millions U.S. dollar
Source: Environmental Leader 2008

Module 5

A Guide to the Green Financial Markets

THE growth of the green financial markets and the introduction of new rating techniques for carbon products are essential for the future of the low-carbon global economy. At the same time, institutional investors are playing an important role in persuading global corporations to disclose their carbon exposure and explain their strategies for cutting down carbon emissions. Carbon risk management and financing is becoming an integral part of the mainstream financial markets. All these developments are good signs for successful transition to the low-carbon global economy.

Green Venture Capital Funds

There are many entrepreneurial opportunities available in carbon trade, clean energy technology, and carbon abatement projects. Among the opportunities is to develop low carbon emissions projects or modify existing projects to cut down GHG emissions. There are also opportunities in developing new carbon abatement technologies and marketing them to other corporations. These activities can be funded by governmental agencies in many developed countries. In the United States, the National Science Foundation (www.nsf.gov), the U.S. National Renewable Energy Laboratory (www.nrel.gov), the American Solar Energy Society (www.ases.org), and various state-funded organizations offer research and seed money for these projects.

The most common financing source for new and existing carbon abatement projects comes from green venture capitalists (GVCs). These are the venture capitalist specializing in green projects. The GVCs provide early-stage funding to projects or companies projecting that they will reduce GHG emissions. The project evaluation process is cumbersome, and often, capital is available only after certain milestones are met. There are dozens of green venture capital funds in the U.S. market. The green venture capital industry has been growing at a rate of 50 percent annually since the Kyoto Protocol went into effect (McCarthy 2006).

In the United States, only 74 green venture capital deals were executed in 2005, with about $600 million changing hands. By 2006, the number of deals grew to 124, and the value of those deals climbed to almost $2.5 billion. Green venture capital got attention in 2007, when the number of deals grew to 222 at a value of approximately $3 billion. By the end of 2008, green venture capital deals were around 300 at a $ 3.85 billion value; however, authentic data arenot available yet (Hodge 2008).

In 2008, the global green venture capital investment was $ 8.4 billion, up 38% from $6.1 billion in 2007, despite the global financial meltdown in the fourth quarter of 2008 (Cleantech Network 2009).For example, there are about half a dozen upstart companies funded by the GVCs developing green electric cars and competing with the global giants such as Toyota and General Motors. Some of these upstarts are already marketing their cars or plan to market cars by 2010, and their names are as follows (Taylor 2008):

- Think Global (Electric commuter car)
- Tesla Motors (Electric sports car)
- Loremo (Diesel / electric fuel-efficient car)
- Mindset (Hybrid coupe)
- Gordon Murray Design (City commuter car)
- Fisker Automotive (Hybrid sports car)

The green capital funds offer an opportunity to investors as well. Green investors can earn good returns and participate in green revolution. To qualify as a green venture capital investor, the investor must be "accredited, " which means that the investor must have a net worth of over one million dollars and should have earned over two hundred thousand dollars in the past two years. The GVCs take an active role to build a corporation, and after making the startup a commercially viable entity, the GVC will walk the corporation through the initial public offering (IPO) or sell the startup corporation to a well-established corporation.

Advantages of Green Venture Capitalists (GVCs)

The GVC investment in carbon abatement projects offers opportunities to project developers as well as investors. The green GVC funds offer at least four advantages to any small potential investor and start-up corporation.

- First, the GVC funders understand the market, and they bring with them the depth of start-up management experience, thus minimizing the failure risk.
- Second, GVC funders understand asset structuring, which means spreading their investment in several technological categories and minimizing risk while improving profitability.
- Third, GVC funders help develop a comprehensive business strategy and provide continuous medium-term financial support.
- Fourth, GVC funders have experience in issuing IPOs or selling corporations to the highest bidder. For a clean energy start-up corporation, GVC funds are a valuable source of funding, and for highnet-worth investors, investing in a well-managed GVC fund offers a high profitability option (Green VC 2008).

Major Green Venture Capital Networks

The green venture capital market consists on relatively small venture capitalists focusing on one sector and large networks investing in multiple sectors and countries. One of the major green venture capital networks is the Cleantech Network (www.cleantechnetwork.com). As of December 2008, Cleantech Network claimed to include over eight thousand investors, six thousand corporations around the world, three thousand five hundred professional service organizations, and over $3 trillion in assets under management. Cleantech Network started in the United States. The network now operates in the EU countries, China, and India. Corporations interested in securing venture capital financing can directly contact Cleantech or attend one of the four annual forums (two in North America, one in the EU, and one in China).

Every green venture capital fund has a unique focus on an industrial sector or geographical region, and green venture capital activities, at times, are geared toward a particular country. For example, in early 2008, Green Venture, a New York-based emission commodity management corporation launched the first ever carbon fund, which has raised approximately $300 million for investment in CDM projects in India and Nepal. The fund will invest in CDM projects as well as the CERs generated by other CDM projects. It was believed that India was underrepresented in global green investment, so this fund should be good news for CDM project developers. The fund is targeting renewable energy projects in these two countries (Carbon Finance, 2008b).

There are also "angel investors" who specialize in carbon abatement projects. Angel investors pool their resources and provide early investment in risky projects that have potential for major success. Angel investors are very selective in their investment decisions. Angel investors may aggressively support carbon projects if the new U.S. administration puts high priority on environmental issues.

Publicly Traded Green Stocks

A large number of green stocks are traded on global financial exchanges, and these stocks offer an opportunity for small investors to participate in green ventures as well. Green stocks represent low-carbon emitting corporations or other private enterprises participating in green investments. These stocks may also offer higher returns compared to the rest of the market. A risk-averse investor may consider investing in green mutual funds. Green mutual funds invest in green stocks from different industries.

Green stocks come from all business sectors and sizes of corporations. As of 2008, there were seventy-two corporations listed on U.S. exchanges that were classified as exclusive clean energy corporations. These clean energy corporations, along with large solar- energy-related corporations, are traded on U.S. exchanges. There is also an OTC market for smaller corporations (with a market cap around or under $100 million). These corporations are traded on the OTC Bulletin Board (www.otcbb.com) or the Pink Sheets (www.pinksheets.com).

Small and large U.S. investors can also take advantage of the international clean energy (green stocks) either by directly investing in foreign stock exchanges or investing in American Depository Receipts (ADRs) or Global Deposit Receipts (GDRs) traded on the U.S. stock exchanges. ADRs and GDRs simply bundle foreign stocks and offer them as a financial instrument on a U.S. equity market. Some of the European and Chinese stocks are cross-listed on U.S. exchanges as well. The green European corporations offer a safer investment opportunity because they have been operating for over a decade and they can provide historical performance report.

For a new investor, it is difficult to pick an individual stock that is profitable. One of the strategies to overcome this obstacle is to invest in a green mutual fund or Exchange Traded Fund (ETF). A green ETF allows investors to invest in the entire sector, thus spreading risk and return over several corporations in that sector. An ETF does not pick a better performing stock; the investor just averages out his or her risk and return. On the other hand, mutual funds try to pick better performing stocks in one or several sectors of green corporations. Mutual funds are managed by professionals who have a history of picking a better stock; thus, mutual funds offer a low-risk opportunity to green investors. Mutual funds do charge a management fee. Some of the larger and better-known U.S. green mutual funds are as follows(Asplund 2008):

- Winslow Green Growth Fund (www.winslowgreen.com)
- New Alternatives Fund (www.newalternativesfund.com)
- Guinness Atkinson Alternative Energy Fund (www.gafunds.com)
- Calvert Global Alternative Energy Fund (www.calvert.com/sri.html)

SUBSIDIZED GREEN PUBLIC FINANCING

In the United States, a wide range of incentives are available from both the federal government and a number of states to support corporations' investment in renewable energy, biofuels, energy efficient processes, and forestation. Almost all U.S. states are offering incentives in various renewable energy sectors. A list of the incentives is available on the U.S. Department of Energy Website (The Green Power Network 2008). For example, the U.S. federal government and fourteen states have been subsidizing industrial wind power generators. These states have set up ambitious goals to generate a portion of their energy from wind farms. For example, wind farm developers are getting an approximateone- to two-cent per kilowatt-hour (kWh) subsidy, in addition to any federal government incentives;Congress periodically renewsthe tariff subsidy. Future investors should pay attention to these programs and consider them in their project financing model.

At the end of 2008, the United States had been producing more wind power than any other country (about 23, 900 megawatts), with Texas leading the pack followed by California, Iowa, and Minnesota (AWEA 2008). These fourteen states have mandated that about one-third of their power should be produced from wind by either 2025 or2030.

Most of the U.S. states, as well as the federal government, have been offering economic incentives and tax breaks to biomass, biodiesel, solar, geothermal, and other renewable energy sectors. Net-metering laws (which govern an energy producer's right to connect to the electricity grid and sell excess power to area utility companies) are enforced in forty-one U.S. states and the District of Columbia, so small investors can take advantage of their state subsidies by installing small, renewable energy projects for residential and small-business use and sell excess power to local utility (Chernova 2008). There are also incentives and federal tax breaks for electric car and hybrid car purchases, as well as for employing energy-saving devices. A green investor must study these incentives before undertaking new investments.

Investors' Demand

Investors are also demanding that corporations disclose their participation in carbon-abatement activities and current carbon risk. Investors are interested in minimizing their potential carbon risk and the risk of financial penalties. One of the most talked about developments is public scrutiny by the Carbon Disclosure Project (CDP), which encourages corporations to reduce GHG emissions. As of 2008, the CDP is a coalition of about three hundred eighty-five institutional investors with $57 trillion under their management. Since 2000, the CDP has made six requests to the largest global corporations to disclose their GHG emissions reduction programs. All six of these survey reports are available to the public at large (CDP 2008). Such a request suggests that multinational corporations and large global corporations are subject to increasing pressure, not just from their governments but also from investment groups, to find a market-based solution to environmental problems and reduce their GHG emissions. The CDP has positively influenced the attitude of global corporations toward carbon management. Environmental consciousness is no longer just considered a "nice" thing to do; instead, corporations' actions have economic implications. The CDP Website is the largest repository of corporate GHG data in the world (CDP 2008).

Banking and Insurance

Commercial banks and other financial service providers are relative latecomers in the carbon finance industry. The carbon finance industry, as a consequence, suffers from a lack of uniform standards and difficulties in risk/return evaluation of carbon projects. Swiss Re and Munich Re were the two pioneer European insurance corporations to get their feet wet in carbon finance projects. Other large insurance corporations, such as AIG and Zurich, have also developed policies to insure against the nondelivery risk of the carbon credits in CDM and JI projects. The insurance industry's involvement in the carbon market is, at best, described as in the early stages of developing carbon trading policies. After suffering major losses in October 2008, future involvement by banking and insurance sectors in the global carbon markets remains to be uncertain. Insurance companies' participation in the global carbon market is related to the following performance risks:

- Regulatory risk of the carbon markets
- Country risk (particularly for CDM projects)
- Technology performance risk
- Counterparty risk
- Liability

Most of the U.S. and international commercial banks have designated a percentage of their loans to green ventures. In addition, these banks require energy efficiency and carbon abatement strategies in all business plans submitted for financing. The rationale for this requirement is that commercial banks wish to protect the project from unexpected carbon emissions compliance costs, which may lower the profitability of the corporation (Asplund 2008). It is also argued that actual implementation of environmental risk management varies among financial institutions, and at times, the financial institutions focus narrowly on the GHG emissions issue while ignoring economic development and sustainability issues (Montgomery 2008). The level of the U.S. and international commercial banks' participation in green projects is a little uncertain after the meltdown of the global financial system in the third quarter of 2008.

QUANTITATIVE MEASURES OF CARBON RISKINESS

Because the carbon strategy of new projects is gaining legitimacy in investment and other financial analysis, there is a need for quantifiable measures of a project's carbon risk. The quantifiable value of the carbon risk can be incorporated in the project's financing and management decisions. Most of the publicly available GHG emissions and management information is extracted from questionnaire responses. Above all, this information is nothing but the subjective opinions of the questionnaire respondents. It is difficult to validate the information and compare the emissions reduction data across industries and countries. In the absence of quantitative measures of the projects, a majority of CDM projects have received certification of just over 70 percent of the desired quantity of carbon credits (Carbon Rating Agency 2008). A detailed analysis of the carbon abatement projects by carbon rating agencies may improve the quantity of carbon credits generated from these projects and, hence, improve their financial health.

Market Initiated Rating Service

The Carbon Rating Agency announced one of the first comprehensive solutions to noncompliance and nondelivery risks in the carbon trading market. In the summer of 2008, the Carbon Rating Agency introduced the Market Initiated Rating Service to evaluate the riskiness of carbon emissions-reduction projects. The third-party evaluation of carbon-reducing projects offers additional information for investors in CDM, JI, and voluntary market projects—areas where occasional bad news may have scared away potential investors and corporations. The Market Initiated Rating Service aims to give carbon market participants access to information about the carboncredit delivery risk profile of a range of CDM, JI, and voluntary projects. The rating agency analyzes projects by focusing on five major categories:

- Project concept
- Project participants
- Project context
- Project implementation
- Emission reduction framework

The prime goal of the Market Initiated Rating Service is to increase market efficiency and transparency and, consequently, enable market participants to profit from improved market function. Furthermore, information on delivery risk and increased transparency reduce the risk of non-approval of new carbon credits-generating projects (Carbon Rating Agency 2008).

Carbon Beta

A leading financial advisory corporation, Innovest, has introduced a quantifiable measure based on the concept of the corporate beta, which is a measure of the riskiness of an individual portfolio vis-à-vis the entire market. Innovest's risk measure is called the Carbon Beta. It is believed that the environmental, social, and governmental parameters that are incorporated in the Carbon Beta analysis have the most impact on corporations' financial and share price performance. The Carbon Beta model addresses three critical factors(Innovest 2008):

- Absolute and relative environment risk exposures for individual corporations
- Corporations' capacity to manage these risks
- Corporations' ability to identify and capture upside commercial opportunities being created as a result of better environment risk management

The **Carbon Beta** platform currently covers over fifteen hundred international corporations in carbon-intense sectors. There are three primary elements for assessing corporate risk and performance. The Carbon Beta analysis incorporates industry sector exposure, corporation-specific carbon analysis, and carbon financials. The Carbon Beta is one example of the quantifiable measure availability in the carbon market.

The Carbon Beta does have its limitations because corporation analysis is focused only on the voluntary carbon disclosure from the corporations, and major corporation analysis is based only on carbon emissions. By focusing exclusively on carbon emissions, the Carbon Beta index disregards the corporation's carbon management efforts and possible improvements, which may turn a corporation's project into an opportunity rather than a risk.

Corporate Rating

Traditional rating agencies (Moody's, Fitch, and S&P) are reluctant to rate projects generating carbon credits; however, these rating agencies at times incorporate carbon management in their evaluation of global corporations. Two leading credit rating agencies (Moody's and S&P) have started incorporating carbon risks and opportunities in their bond rating activities of global corporations.

Another measure of a corporation's involvement in environmental issues is the Dow Jones Sustainability Index (DJSI). The DJSI is an index available to rank global corporations based on their social responsibility. Carbon management is just one of the elements of the DJSI and bond ratings. The DJSI offers investors professional benchmarks and an investment universe for active and passive sustainability portfolios. This index also provides investors with a platform to encourage corporate progress toward sustainability and long-term business success. In that context, a growing number of corporations desire to be ranked favorably by the DJSI and hope that favorable ranking will highlight their sustainability credentials (DJSI 2008).

Module 6

Global Carbon Industry Leaders

THIS module discusses the role of several multilateral institutions in the growth of the global carbon industry. Carbon credits trading and the need to reduce carbon emissions are slowly being accepted as new realities of global business. A number of institutions played crucial roles in developing the global carbon market. The World Bank Group, the World Resource Institute, the Institutional Investor's Group on Climate Change, and the Carbon Disclosure Project are some of the pioneers in developing the global carbon industry.

THE WORLD BANK GROUP

The World Bank Group has been a proactive player in the carbon finance market. One of the organizational units exclusively involved in carbon projects is the World Bank Carbon Finance Unit (CFU). Another initiative by the World Bank is to set up the Strategic Framework on Climate Change and Development (SFCCD), which is a partnership among member countries to tackle global environmental problems (Carbon Finance Unit 2008). In 2002, only the World Bank and the Dutch government were involved in carbon purchasing programs, but as a result of the World Bank's leadership, there are close to one hundred public and private carbon funds in the market (The World Bank 2008).

1. **The Carbon Finance Unit**
   The Carbon Finance Unit (CFU) is funded by contributions by global corporations and governments who are members of the Organization for Economic Cooperation and Development (OECD). Funds are used for the purpose of purchasing project-based greenhouse gas (GHG) emissions reductions in developing countries (CDM projects) and economies

in transition (JI projects). The emissions reductions are purchased through one of the CFU's carbon funds on behalf of the contributor, for CDM projects or for Joint JI projects. The list of the CFU's carbon funds and facilities is presented in Table 6.1.

Unlike other World Bank development finance units, theses carbon funds and facilities do not lend funds to carbon projects. Rather, they offer contracts to purchase verified CO_{2e} emissions reductions in a commercial transaction, paying for them annually or periodically. The purchase of carbon credits by these funds and facilities, as a result of emissions reductions, add an additional revenue stream in hard currency and improve bankability of these projects. This reliable revenue stream improves projects' cash flow and reduces the risks for commercial lenders. Thus, carbon finance provides a means of leveraging new private and public investments into projects that reduce GHG emissions, thereby mitigating climate change while contributing to sustainable development. Without the additional revenue generated by carbon credits, these projects might not be possible.

Another contribution of the World Bank's CFU is to strengthen and streamline the Kyoto Protocol mechanism. The Policy and Methodology Team in the CFU systematically observes the CDM regulatory process and contributes to rulemaking for the CDM by interpreting regulatory decisions, providing input, and developing new methodologies. Through these actions, the Policy and Methodology Team bridges the gap between general guidelines and practical methodologies for carbon projects. The CFU also prepares policy and position papers and takes an active role in initiating research and studies on methodological and policy issues related to the CDM. Almost all the CFU publications are available to global corporations and investors (The World Bank 2008).

Table 6.1

World Bank Carbon Funds and Facilities

	Name of Fund/Facility	Key Features
1	Prototype Carbon Fund	Operating since April 2000, this is the first carbon fund. This fund is a partnership between seventeen corporations and six governments. The fund's total capital is about $180 million.
2	BioCarbon Fund	The BioCarbon Fund is a public and private partnership promoting biodiversity and conservation. The fund purchases carbon credits from land use and forestry projects.
3	Community Development Carbon Fund	This fund started in March 2003 with the objective of uplifting poor developing countries and reducing their carbon emissions. The fund operates with the assistance of public and private sectors.
4	Italian Carbon Fund	This fund started in fall 2003. The fund purchases carbon credits from developing countries for Italian public and private sectors.
5	The Netherlands CDM Facility	This facility was established through a cooperation of the Netherlands and the World Bank with the sole purpose of purchasing carbon credits from CDM projects in developing countries.
6	The Netherlands European Carbon Facility	This agreement between the Netherlands and the International Finance Corporation (IFC) was signed in August 2004. The fund purchases carbon credits on the behalf of the government only from JI projects located in economies in transition.

7	Danish Carbon Fund	The Danish Carbon Fund is an example of a public and private sector cooperation to assist Danish businesses. The fund was set up in January 2005.
8	Spanish Carbon Fund	This fund invests in JI and CDM projects and promotes cleaner technologies and sustainable development. The fund was set up in 2004 with the cooperation of the Spanish government and the World Bank.
9	Umbrella Carbon Facility	This is an aggregating facility to pool funds from the International Bank for Reconstruction & Development- (IBRD) managed carbon funds. The facility invests in large projects because the total funds available are $719 million.
10	Carbon Fund for Europe	This is a trust fund established by the World Bank, in cooperation with the European Investment Bank (EIB). Both prestigious institutions bring their expertise to purchase carbon credits from CDM or JI projects.
11	Forest Carbon Partnership Facility	This facility devises policies and offers economic assistance in activities that would reduce emissions from deforestation and degradation. As of 2008, about forty countries had expressed an interest in participating.

Source: Carbon Finance Unit 2008

2. The World Bank Plan for Forests

The World Bank has established a number of funds and programs primarily to assist medium- and low-income countries in their GHG reduction efforts and save their forests. The Forest Carbon Partnership Facility (FCPF) and the BioCarbon Fund are the two major initiatives of the World Bank. The financing model of these two funds also highlights the importance of public and private sectors cooperation in saving the forests and reducing GHG emissions (Jacquot 2008).

Table 6.2 lists the GHG emissions of various economic sectors in the member countries of the World Bank Group (International Bank for Reconstruction and Development and the International Development Association). Land-use change and forestry account for 32 percent of GHG emissions. The World Bank Group realizes the importance of land-use and forestry sector in their efforts to manage GHG emissions. Electricity and heat generation contribute another 20 percent of emissions. Agriculture and transportation are the third and fourth major contributors of GHG emissions.

Table 6.2

GHG Emissions by Sectors (IBRD and IDA Members)

Economic Sectors	Percentage of Total Emissions
Land-use change and forestry	32%
Electricity and heat	20%
Agriculture	15%
Industry	11%
Other sectors	11%
Transportation	6%
Waste	3%

Source: www.worldbank.org

3. The Carbon Partnership Facility

In 2008, the World Bank set up the Carbon Partnership Facility (CPF) with an initial funding of $500 million. The main objective of the CPF is to provide financial assistance to carbon reduction projects. The CPF will purchase carbon credits to be generated beyond 2012, unlike other funds/facilities, which have a short-term assistance horizon. The facility promotes long-term and somewhat risky investment programs for sustainable development. Some of the targeted sectors are renewable energy, fuel substitution in power generation and industry, waste management systems, carbon capture and storage, and a myriad of energy-efficient industrial systems.

4. **The Clean Energy Investment Framework**

    The objective of the Clean Energy Investment Framework (CEIF) is to facilitate development of low-carbon emitting energy sources. It is believed that increased energy sources are crucial for economic development and that a country can develop new sources of energy while reducing its carbon emissions. There are multiple programs available to countries, including loans, grants, and finance guarantees for the clean energy sector.

5. **Climate Investment Funds**

    Another initiative of the World Bank Group includes the Climate Investment Funds (CIFs), which were approved by the World Bank as in early 2008. CIFs were set up to promote international cooperation on climate change and to support the Bali Action Plan of December 2007. The CIFs are an important new source of interim funding through which the multilateral development banks (MDBs) will provide additional grants and concessional financing to developing countries to address urgent climate-related challenges. The CIFs will enable a dynamic partnership between the MDBs and developing countries to undertake investments that achieve a country's development goals while choosing a low-carbon development path (The World Bank 2008).

The two major climate funds are the Clean Technology Fund and the Strategic Carbon Fund.

- The Clean Technology Fund will complement existing green financing mechanisms. The Clean Technology Fund is available for private and public sectors in their efforts to mitigate GHG emissions and transform participating economies into low-carbon economies. There are wide arrays of financial instruments available, including grants, subsidized loans, bridge financing, and loan guarantees.
- The Strategic Climate Fund promotes programs that have potential to bring lasting impacts on multiple sectors within a country. The

fund also targets sustainable forestry and agriculture projects and shares knowledge among the MDBs on climate change.

THE WORLD RESOURCE INSTITUTE

The World Resource Institute (WRI) is an environmental think tank involved in research and analysis of global sustainability issues and making policy recommendations. The WRI publishes numerous reports and case studies to guide profit and not-for-profit entities to manage their carbon emissions. WRI studies have also focused on the forestry industry as one of the solutions to global environmental problems. The WRI mission statement focuses on educating citizens and corporations on ways to protect the environment, offering suggestions about market-based solutions to tackle global environmental problems, helping industries to set up standards, and preserving forests (WRI 2008). The WRI head office is located in Washington DC. The WRI was set up in June 1982, and it was one of the earliest organizations to target global environmental problems.

One of the significant accomplishments of the WRI is development of the Carbon Value Analysis Tool (CVAT). CVAT is a screening tool that helps corporations makingenergy-related investment decisions that reduce GHG emissions. The corporations and their investors can test the sensitivity of an energy-related project's internal rate of return (IRR) to the quantity of GHG emissions reductions. The corporation or the investors can establish a relationship between GHG emissions reductions and profitability of the project. CVAT integrates this value into traditional financial analysis by assigning a market price, either actual or projected, to carbon emissions reductions. The World Resource Institute screening tool (CVAT) estimates reductions of direct and indirect emissions using standards developed by the GHG Protocol Initiative (WRI 2008).

THE INSTITUTIONAL INVESTOR'S GROUP ON CLIMATE CHANGE

In 2001, eleven institutional investors formed the Institutional Investor's Group on Climate Change (IIGCC), and it currently targets three key industries: power generation, aviation, and construction.As of June 2008, IIGCC was made up of about fifty major European institutional investors and pension funds that emphasize environmental issues in their long-term investment decisions. The objective of the IIGCC is to encourage corporations to disclose their carbon emissions and adopt socially responsible GHG emission strategies as prerequisites of future financing (IIGCC 2008).

According to the official Website of the group, the purpose of the IIGCC is to promote a better understanding of the implications of climate change amongst group members and other institutional investors. Furthermore, the IIGCC encourages corporations and markets that it invests in to address any material risks or opportunities to their businesses associated with climate change or a shift to a lower carbon economy. This is one of the earliest, and perhaps the first, organized efforts by institutional investors to spotlight carbon issues as an investment variable.

The Carbon Disclosure Project

In December 2000, thirty-five institutional investors set up the Carbon Disclosure Project (CDP). The CDP is an impartial not-for-profit entity with the objective of creating "a lasting relationship between shareholders and corporations regarding the implications for shareholder value and commercial operations presented by climate change" (CDP 2008). The purpose of the project was to make inquiries and to encourage large global corporations to adopt environment-friendly measures in their business operations. These institutional investors were concerned about unexpected future risks as a result of investing in polluting industries and exposing themselves to new regulations, which might negatively influence profitability. There has been a sequence of efforts by the CDP, each with increasing scope and reaching out to the more global corporations, to collect data on the low-carbon initiatives by the surveyed corporations.

In 2002, these 35 institutional investors managing $4.5 trillion requested the largest global 500 corporations to disclose their susceptibility to climate risk. This initiative was called the first Carbon Disclosure Project. Overall findings were not very encouraging because corporations were not paying attention to carbon exposure and because the response rate was also lower than expected. The second Carbon Disclosure Project was launched in 2004. A major achievement was the third Carbon Disclosure Project, which was launched in 2005 and included 143 institutional investors managing $21 trillion. The number of concerned institutional investors increased by more than fourfold in three years, and corporations could no longer ignore the questionnaire sent by the CDP. The response rate to the questionnaire improved from 78 percent in 2002 to 89 percent in 2005; at the same time, the fullycompleted questionnaire response rate increased from 47 percent to 71 percent (CDP 2008).

In early 2006, the group made its fourth request to 1, 933 of the world's largest publicly traded corporations to disclose their investments in GHG reduction projects (CDP 2008). The 2006 list included Fortune 500 corporations, 258 of the world's largest electric utilities, 300 of the largest GHG emitters in Canada, 200 of the largest corporations in Germany, 150 of the largest corporations in Japan, 150 of the largest corporations in Australia and New Zealand, 120 of the largest corporations in France, 100 of the largest corporations in the United Kingdom, 50 of the largest corporations in Brazil, and 40 of the largest corporations in Asia outside of Japan. All the 1, 933 corporations were requested to complete the questionnaire within four months. The fifth Carbon Disclosure Project was launched in 2007, on behalf of 315 institutional investors managing $41 trillion assets.

In 2008, the total asset value of the 385 CDP-related institutional investors increased to $57 trillion (a thirteen-fold increase since 2002). The sixth Carbon Disclosure Project was launched in 2008 and targeted about 3, 250 large corporations from the United States, Europe, Latin America, Australia, South Africa, and Asia, including China. This is the most comprehensive study of large global corporations' attitudes and policies towards GHG emissions issues. The questionnaire included queries under the following categories:

- Risk and opportunities
- GHG emissions accounting
- Additional GHG emissions accounting
- Corporate performance
- Corporate governance

The results of the project are publicly available. Based on these survey results, the CDP has created the Carbon Disclosure Leadership Index (CDLI). The index includes the corporations with the highest scores in the two categories of the carbon-intensive sectors and the non-carbon-intensive sectors and provides a valuable perspective on the range and quality of responses to the CDP's questionnaire. The sixth CDP survey report is also available for various subgroups (Global 500, S&P 500, FTSE 350).

The list of the corporations and groups surveyed for the sixth CDP project (May 2008) is found in Table 6.3.

Table 6.3

Corporations Surveyed for the Sixth CDP Project

	Origin of Corporations	**Number of Largest Corporations***	**Affiliation of Corporations**
1	Global Corporations	500	Financial Times 500
2	The United States	500	S&P 500
3	The United Kingdom	350	FTSE 350
4	Australia	200	ASX 200
5	Canada	200	
6	Germany	200	
7	India	200	
8	Nordic Countries	193	Nutek, Folksam, KLP
9	Japan	150	
10	France	120	SBF 120
11	China	100	
12	South Africa	100	FTSE/JSE 100
13	Switzerland	100	
14	Global Transport	100	
15	Brazil	75	Sao Paolo Exchange
16	Asia excluding China, Japan, and South Korea	50	
17	South Korea	50	
18	Netherlands	50	
19	New Zealand	50	NZX 50

20	Italy	40	
21	Latin America	40	S&P Latin America
22	Spain	35	IBEX 35

*Large corporations based on market capitalization

Module 7

Importance of Carbon Footprints

THE Carbon footprint of a product or service is a measure of GHGs emitted from the production and use of the product or service over a prespecified period of time. The overall carbon footprint of the human race is a measure of the global amount of GHGs emitted by human activity (Carbon Footprint 2008). All corporations and institutions must develop systems to calculate and reduce their carbon footprints. The low-carbon footprints or attempts to lower carbon footprints of a product or service can be used as a marketing slogan as well as for generating carbon credits for cash. One of the problems is a lack of uniform standards for carbon footprints calculation. Corporations are counting their carbon footprint differently, which creates confusion; thus, uniform standards must be developed and adopted by global corporations (Ball 2008a).

An accurate estimate of a corporation's carbon footprints will determine the necessary level of carbon emissions reductions in years to come. Carbon footprints cannot be estimated by an external consultant alone. The consultant may provide valuable assistance, but the concerned persons in the corporation, sector, or country must gather consistent carbon emissions data about its daily operation and try to understand the sources of that data.

The most crucial component of any market-based carbon reduction program is to establish a baseline from which a corporation, sector, or country measures future carbon emissions reductions. A low baseline may grant a corporation, sector, or country more valuable carbon credits, and similarly, a higher baseline may not be as beneficial to the corporation, sector, or country. In a voluntary market or a pre-compliance market, such as the U.S. market in 2009, corporations must formulate strategies to accurately estimate their current carbon footprints.

Greenhouse Gas Protocols

A greenhouse gas protocol defines and explains accounting rules and standards to estimate GHG emissions. The protocol provides the accounting framework for nearly every GHG standard and program in the world. These protocols also offer developing countries an internationally accepted tool to help their businesses to offer credible GHG data and their governments to make informed decisions about climate change.

Selection of the appropriate industry and region specific protocol is critical for a credible carbon footprint estimation process. There are at least two dominant GHG estimation protocols.

- The first GHG protocol was developed by the cooperation of the World Resource Institute (WRI) and the World Business Council of Sustainable Development (WBCSD). The first GHG protocol was published in 2001. The most widely used protocol is the first GHG protocol, which contains detailed calculations and accounting standards. More than one thousand leading global institutions and 63 percent of U.S. Fortune 500 corporations have been employing this GHG protocol in their estimations of GHG emissions.
- The second protocol is called the GHG Protocol Corporate Standard. The International Organization for Standardization (ISO) developed this standard in 2006 as the basis for its *ISO 14064-I standard*. As of December 2007, the ISO, WRI, and WBCSD had signed a memorandum of understanding (MOU) to promote the first and second GHG protocols. The United Nations and leading industrial countries often provide their input to modify these two GHG protocols (GHGPI 2008).

Greenhouse Gas Accounting Principles

GHG accounting and reporting principles are new and still evolving. First, these principles must be in accordance with the accounting principles of the country of operation. Asecond successful ingredient of these accounting principles is acceptability to all stakeholders. Third, the GHG emissions estimates should be complete and account for all emissions sources and activities and should keep an emissions inventory for all subsidiaries of the corporation in all geographic locations. The detailed GHG emissions data is necessary in developing region and subsidiary-specific strategies of reducing GHG emissions. Fourth, the accuracy of the GHG emissions estimates is equally important for internal and external decision making. The accounting practices should be transparent and address all relevant issues in a factual and coherent manner. Fifth, the accounting practices should employ consistent methodologies to allow meaningful comparisons of GHG emissions over time. Above all, the results should be credible, which implies that GHG emissions are neither over- nor underestimated (GHGPI 2008).

Inventory Design and Business Goals

The objective of the GHG inventory design should not only be to estimate GHG emissions but also to offer a comprehensive tool to formulate future low-carbon investment strategies. The GHG inventory design should have a long-term objective, and the emissions data should be easy to understand and analyze while deciding among various investment opportunities. A successful GHG inventory design system should also automate assembly of data and analyze changes in GHG emissions under various real and hypothetical scenarios. Above all, GHG inventory design should automate baseline projections at the end of each year.

Operational and Organizational Boundaries

The determination of the sphere of responsibility for a corporation's GHG emissions is a critical issue. In other words, the boundaries of a corporation's responsibilities need to be specified. There are at least two methods to define the boundary of the corporation's GHG emissions responsibility. First, the equity share method permits accounting of GHG emissions according to the corporation's share of equity in the emitting organization or operation. This means that a 51 percent share in a venture will make the corporation responsible for only 51 percent of the GHG emissions. Second, the control level method suggests that a controlling corporation, even if the equity share is only 51 percent, should be responsible for 100 percent of the GHG emissions. This gap in the amount of GHG reduction responsibility assumed by a controlling corporation (depending on whether the corporation adopts the equity share method or the control level method) needs to be defined before the start of the carbon footprinting process. There may be industry- or country-specific practices, which may influence the selection of the equity share or control level methods for estimating carbon footprint (GHGPI 2008).

Scope of Emissions

The scope of the corporation's emissions is another critical factor in accurate accounting of GHG emissions.

- Scope 1 emissions are from sources that are owned or directly controlled by the corporation. Some of activities included in the scope 1 emissions are transportation of material and employees, at site production of electricity and heat, and industrial processing methods. It is much easier to estimate and control scope 1 activities than it is to estimate and control scope 2 and 3 activities.
- Scope 2 emissions are from purchased heat and electricity, so knowledge about the GHG emissions of the utility corporations is equally important. For example, the corporation may have two options for electricity purchase. The first option is electricity generated from coal, and the second option is electricity generated from wind turbines. It is obvious from a carbon footprint standpoint that the second option is preferable because purchase of electricity from a wind farm will lower the scope 2 emissions estimate of the corporation. In cases of multiple sources of electricity and heat generation by the utility corporation, the weighted average of GHG emissions will be estimated.
- Scope 3 emissions encompass all indirect GHG emissions, which can be somewhat difficult to estimate and manage. Scope 3 emissions of a corporation are actually scope 1 emissions of a different corporation that provide some of the services. Scope 3 emissions include business travel, outsourced business activities, production of imported material, emissions from a corporation's waste and final disposal of products, and employee emissions when commuting to work. For example, an airline is responsible for its own GHG emissions (scope 1 for the airline), but a corporation using an airline (scope 3 emissions) has the choice to select an airline that emits low GHGs or cut down air travel. At present, corporations have a limited choice to control their scope 3 emissions because detailed GHG emissions data is not widely available.

The purposes of the segregation of the scopes of GHG emissions areto not double count emissions data and to hold organizations responsible for their business decisions' impact on GHG emissions. Scopes 1 and 2 are mandatory under both GHG protocols. There is a risk of some overlap between a corporation's indirect emissions (scope 2 and 3) and direct emissions (scope 1); thus some double accounting is inevitable (GHGPI 2008).

Timeline and Cost

The process of designing the GHG emissions reduction processes, setting up reporting protocols, testing and implementing the GHG emissions process, and finally, quantifying a corporation's carbon footprint is costly and time-consuming. The timeline and the cost depend on the organizational structure of the corporation and timely availability of relevant data. An external consultant cannot calculate carbon footprints without the assistance of a trained team of managers from within the corporation. Corporations must internalize their GHG emissions estimation process to shorten the time horizon and lower the cost of estimating their carbon footprints.

Benefits of Carbon Footprint Data Collection

Developing a GHG inventory system can facilitate the identification of emissions sources for internal GHG reductions and can possibly reduce the cost of business operation. In the process of becoming the first carbon neutral business in their industry, the Dell Corporation is saving about $3 million annually through facilities improvements and a global power-management initiative (Dell 2008). The real benefit of the carbon footprint process is that the process adds value to the corporation and eliminates any unexpected future business risk due to changes in the carbon compliance regime or pressures from external stakeholders.

Carbon footprints information provides consumers additional information in purchasing a product and reducing their own carbon footprint. For example, a study of several of Toyota's models revealed that the Prius has a lifetime carbon footprints of 44 metric tons. The Corolla, the Camry, and the 4Runner have lifetime carbon footprints of 64 metric tons, 95 metric tons, and 118 metric tons, respectively (Ball 2008a). Consumers may take into account carbon footprints of the Toyota models in making their selection. Many corporations are considering putting carbon labels on their products, in an attempt to differentiate themselves from corporations offering high-carbon products.

The United Kingdom-based organization, Carbon Trust, has been using the carbon reduction label to help consumers understand the carbon footprints of the products and services they use. The purpose of these labels is to help consumers make informed decisions about their purchases (Carbon Trust 2008). The carbon labeling system should be accompanied by an education program of how companies and individuals can use and read the label. Just like nutritional labeling, carbon labeling must be easy for consumers to understand. Now more than ever, corporations are establishing green programs; therefore, a carbon label system would be a nice addition to such programs as it would allow corporations to show that they, indeed, are working toward a low-carbon future. Corporations might be able to charge a premium price for the low-carbon products to green buyers.

Module 8

Financial and Career Opportunities in the Carbon Market

A new economy revolving around the goal of reducing carbon emissions not only improves quality of life for the current generation and generations to come, it also creates financial and career opportunities for citizens. The opportunities mentioned in this section offer financial rewards and self-satisfaction to the people involved in carbon reduction projects and activities. We can be involved in the growing carbon market to find purpose in our lives and improve quality of life for other citizens.

Carbon market education and the consultancy market will expand if and when the U.S. government starts its cap-and-trade program and/or begins participating in a post-Kyoto mechanism along with China and other developing countries. This is a highly likely scenario after the expiration of the current Kyoto mechanism in 2012 (Point Carbon 2008). There are already three regional cap-and-trade initiatives in the United States, so corporations operating within these regions must start planning for compliance and are advised to hire consultants trained in carbon emissions reduction strategies. There is, however, a great need for carbon consultants with better understanding of the carbon regulations and market. In some cases, these carbon-focused consulting groups may help arrange financing by selling projected carbon credits or tapping into green public and private financing sources. In Europe, there are a number of consulting groups that advise CDM and JI project developers from the beginning and help them to generate, register, and validate carbon credits production throughout the life of a project.

1. **Become a Corporate or Community Educator**

   Changing environmental regulations at the local, national, and global levels have created a need for educational services for corporations and local communities. These educators have the bourgeoning task of educating their corporations and communities on how to meet the challenges of functioning in the new carbon reduction environment. These educators must also understand the technical details of carbon abatement projects and identify the costs and benefits of compliance within the framework of the ever-changing environmental regulations.

   Most small and medium-sized corporations, as well as all the corporations in carbon-intense industries (utilities, oil and gas, mining, cement, steel, etc.), need in-house experts helping them to devise business and financial strategies to cut carbon emissions and to become environmentally friendly business entities. These in-house experts will be in high demand in the United States and around the world. Corporate educators can offer intense training to the designated in-house experts. Community educators can sponsor day-long workshops inviting city and state government leaders and local business leaders from their regions. These educational programs should be future oriented and tailor-made to the needs of the particular group. There are monetary rewards for corporate and community educators who take the lead and establish their names at early stage.

   The aforementioned suggestions are for all interested persons, not solely aspiring community and business educators. All citizens can benefit from educating themselves about the new low-carbon economy. Interested persons may choose to attend workshops organized by area universities and government agencies. In addition, a select few universities offer courses that focus on environmental finance issues. Above all, interested persons should devote five to ten hours a month reading about new developments in carbon market. Suggested sources of information are the carbon industry leader's Websites (see Module 6). Business and engineering schools' curricula may also need revision in light of the growth of low-carbon global economy and the rapid changes in GHG emissions reduction requirements in various regions of the world.

2. **Become a Carbon Management Consultant**

   The global carbon credits market is in the early stages of development, and this is particularly true for the United States. Corporations and managers need training and guidance to generate and verify carbon credits from their current and future projects. There are few reputable carbon management consulting groups serving large corporations, but these seminars are expensive (about one thousanddollars per day), and they only reach a limited number of managers. These big-name carbon consulting groups cannot satisfy demands for all small to midsized corporations. There is, potentially, a need for thousands of carbon management consultants targeting certain industries, geographical regions, or sizes of corporations. There is a need for low-cost and practical training programs to prepare small- to medium-sized U.S. corporations to benefit from the growing carbon market. A carbon-reduction project development and credits certification consultant can train interested corporations to function in the new carbon environment.

These consultants can guide and educate their clients about the new reality of carbon management as an important factor examined by potential investors, green consumers, and regulatory authorities. Furthermore, consultants can help their clients to estimate current carbon footprints, which will be used as a "baseline" for future carbon compliance or voluntary carbon reduction initiatives. These carbon consultants will likely be in high demand if they show their clients that timely carbon abatement business strategies may not only reduce cost of compliance but may be financially rewarding to their corporations.

Another consulting activity is assisting their clients to develop carbon trading strategies within their country or between two countries. Consultants should have an in-depth knowledge of the CDM market, JI projects, EU ETS requirements, and key voluntary market regulations. Furthermore, the consultants should understand the operation of global climate exchanges and the OTC carbon finance markets. Carbon consultants can help their clients to reduce their carbon footprints, set up CDM projects in developing countries, and purchase low-cost carbon credits to offset carbon emissions reductions in the United States. There are already several project developers in the CDM market, and the market has enormous potential for growth. These carbon-focused consultants can also periodically offer workshops for their clients to better prepare them to meet new challenges brought on by carbon reduction concerns. In this respect, carbon consultants also assume the role of carbon educators. People interested in becoming carbon market consultants may like to attend workshops organized by area universities and government agencies. Above all, interested persons should devote a few hours a month to reading about new developments in the carbon market.

3. **Start Up a Green Business and Generate Carbon Credits for Cash**

    The proposed U.S. carbon cap-and-trade system will allow limited trade of carbon credits from the voluntary or CDM markets in developing countries, but it is expected that there is no or minimum limit on carbon credits offset within the United States. It is also known that industries such as utilities and manufacturing cannot redesign their processes overnight and thus cannot immediately meet their carbon reduction targets; thus, offsetting may be the only short-term solution for them.

    Any GHG emissions reduction requirement in the United States offers a real economic incentive for entrepreneurs to start environmental friendly businesses (renewable energy, energy efficiency, forestry, etc.) that generate carbon credits, which can be sold to noncompliance corporations for cash. These cash-generating green businesses can be set up in the United States or in developing countries. On the other hand, by employing better carbon management strategies and investing in energy-efficient systems, even heavily polluting corporations can find themselves in compliance and in a position to develop excess carbon credits for cash trade. Carbon emissions reduction strategies and new green projects can become a source of new financing.

4. **Make Use of Subsidies Offered by Local and National Governments**

    In the United States and around the world, regional and federal governments are offering financial incentives for carbon abatement and environmental friendly projects. These incentives result from global governments' efforts to solve their environmental problems because these governments are forced to comply with an international treaty or protocol. These incentives are usually for a limited time period, but they can improve cash flow during the early days of a project. For example, in the United States, the wind energy industry receives federal subsidies on its green power output, and about fourteen U.S. states also offer additional subsidies and taxbreaks on green projects. Many states also offer subsidies training and education for interested people and corporations. TheAmerican Recovery and Reinvestment Act (signed by President Obama on February 17, 2009) has set aside more than $ 42 billion for green industries, energy related improvements to homes and office buildings, loan guarantees and direct subsidy for renewable energy projects, and grants for next-generation electric car battery. A significant portion of $ 42 billion is allocated for rebates to buy energy efficient appliances and tax credit for the purchase of energy efficient residential air conditioners, heat pumps or furnaces. A detailed analysis of subsidies offered by the U.S. and global governments must be a part of all strategic decisions made by corporations. Interested persons must periodically check with the U.S. and global governments about availability of new subsidies and incentives.

5. **Investigate Green Venture Capital and Financing**
   A large number of venture capital firms, commercial banks, and government-supported financing enterprises are targeting new, environment-friendly projects and carbon-abatement processes in existing projects. The principals of these venture funds believe that green investments will reap higher profits in the carbon-sensitive regulatory environment. Entrepreneurs and corporations who had an early start should receive financing at easier terms and should also be able to trade their excess carbon credits for financial gain. Green financing may not be easily available if more and more corporations start applying for this capital. For details, refer to module 5.

6. **Become a Carbon Market Broker or Trader**
   The carbon credit market is expected to be one of the largest commodity markets, if not the largest commodity market, by 2025 (Harvey 2008). There is a consensus that global carbon trading will dominate other financial market transactions. There is a growing OTC market for carbon products, and most of the conventional commodity exchanges, including the New York Commodity Exchange, are adding new carbon products. There is a need for commodity brokers and traders to better acquaint themselves with the ever-evolving carbon market so they can serve their clients better. Traders and brokers will be financially rewarded if they understand global carbon finance mechanism. At the same time, more

brokers and traders will be needed to work on new climate exchanges and on functioning commodity exchanges. Thus, the growth of the U.S. and global carbon market opens doors for more qualified carbon market brokers and traders. It is a good idea for interested persons to take a college course, attend seminars, and, at minimum, read about new developments in carbon trade and carbon finance.

7. **Become an Information Services Provider**

There is a potential for several startup entities to streamline global carbon market information gathering and dissemination, perhaps specializing in a particular business sector or country. Some of the leading information providers discussed in this section (Point Carbon, CDP, etc.) provide broad and general information, ignoring the details of industries and countries. This is another area of enormous growth in the current and post Kyoto-2012 global carbon environment, and after the introduction of a U.S. carbon cap-and-trade mechanism, the carbon market is expected to grow twotothree times the present level. This larger market will need better information gathering and dissemination services than those that currently exist.

Carbon market financial data (spot price, future price, and volume) is available through global climate exchanges and carbon brokers. The real challenge is to collect and disseminate requested micro information about EU ETS, CDM contracts, JI contracts, and voluntary markets to interested groups. Furthermore, analysis of all carbon markets, recent developments in various carbon markets and regulatory issues need to be provided in a timely manner so investors can make informed decisions. Some of the available data and information sources are mentioned in this section. The leading source of broad and general carbon market news and data is Norway-based Point Carbon (www.pointcarbon.com). Point Carbon lists CDM and JI projects in each country, along with daily price charts, global governments' policies, industry-specific news, and news pertaining to corporations. Another source of broad and general information is the Carbon Disclosure Project (CDP). CDP'sWebsite is perhaps the most comprehensive source of carbon and environmental policy-related information for about three thousand global corporations. Corporations' GHG emissions data and policies are available for public viewing. Most of the reports of past CDP surveys are also available;

however, information on the CDP Website is at least a few months old. Other major business and political news sources such as Reuters, Dow Jones Newswires, Platts, and Argus also provide different type of carbon market news.

8. **Become an Accounting Advisor for Green Projects**

    The carbon finance market is still evolving and is heavily influenced by national and international regulatory regimes. Because of these factors, accounting practices dictating how to treat carbon credits (or carbon contracts) in financial statements are not fully developed. Accounting practices in the carbon market are based on an analysis of case law and recommendations from national governments. Carbon contract is a legitimate asset class and should be viewed as a key factor in project-financing decisions. Carbon exposure can be treated as an asset or liability in cash flow-analysis and project financing. Carbon management may be a source of financial gain or expense, return or risk, depending on the treatment of carbon emissions in a project. There is also a need for accountants developing procedures to calculate carbon footprints.

    Accountants and financial analysts who specialize in global carbon finance issues may find themselves in high demand in the near future. Attending pertinent workshops and undertaking self-study are the only two options to acquire expertise in green accounting.

9. **Become a Legal Advisor for Green Projects**

    Carbon trade and finance transactions usually involve parties from two different countries trading an intangible commodity; thus, transactions require well-established legal procedures to transfer the title of the carbon contract and to protect each party. Many international law firms have developed expertise in the carbon market, but there is still huge need for better-qualified lawyers offering their services to green project development, carbon market investors, and regulatory agencies. Interested lawyers and law firms can specialize in a particular business sector or country to better serve their clients.Legal experts can benefit from new information available on several professional organizations'Websites, and by attending conferences and workshops sponsored by professional organizations such as the HG.org (www.hg.org) and Environment Law Institute (www.eli.org).

Module 9

The Future of the Global Carbon Market

THE global carbon market could be worth more than three trillion dollars in 2020, if the United States government participates in a carbon cap-and-trade system to limit GHG emissions or participates in an international agreement in the post-Kyoto 2012 regime. The future size of the global carbon market is based on the assumption that the proposed cap-and-trade system in the United States or international agreement will mandate a 25 percent reduction of the United States' GHG emissions below 1990 emissions levels. The global carbon trade crossed $60 billion in 2007, a growth of 80 percent over 2006. In terms of volume, 2.1 billion tons of CO_{2e} credits were traded, a hefty 64 percent increase over 2006. Despite the global financial meltdown in October 2008, carbon trading reached to around $118 billion in 2008. This reflects a rise in both the volume of carbon emissions transacted and the value of carbon credits. It is also estimated that growth in carbon marketswill continue, reaching $150 billion in 2009 (New Carbon Finance 2009). An accurate estimate of the future carbon market, however, is subject to future carbon targets and the participation of major polluting countries in any future global treaty (Point Carbon, June 2008).

As a result of many bold initiatives by European countries, the global carbon market is growing at a rapid pace. The market in GHG emissions could outstrip the conventional commodities markets to become the biggest traded commodity market. Bart Chilton, commissioner of the Commodities Futures Trading Commission (CTFC) said, "The potential size and scope of a structured carbon emissions market in the U.S. is unequivocally vast. It is certainly possible that the emissions markets could overtake all other commodity markets" (Harvey 2008). The probable exponential growth of the carbon market is attributed to the seriousness of major countries about tackling global environmental problems and reducing current dependence on fossil fuels. Developed and developing countries have to make major cuts in their fossil fuels consumptions by employing green technology and different energy-efficient methodologies. In the process, they can generate carbon credits worth billions of dollars.

An attempt to reduce global GHG emissions reduction cannot be successful without including most of the major polluting nations. If the status quo is maintained, particularly as developing countries such as China and India continue heavy coal use to fuel their economic growth, man-made emissions of GHGs from the use of fossil fuels is likely to rise over 50 percent by 2030, and climate control may become a difficult task (Point Carbon 2008).

The Post-Kyoto Protocol Global Carbon Regime

Participants in the global carbon regime are monitoring two key factors. First, they are closely observing developments leading to the new carbon regime after the expiration of the Kyoto Protocol in 2012. Second, they are exploring the possibility of adding additional countries targeting carbon reduction.

A number of countries agreed to the general framework of the post-Kyoto Protocol2012 carbon regime in the Bali Mandate signed in December 2007. Members of the United Nations Framework Convention on Climate Change (UNFCCC) including several European countries (advocates of the Kyoto Protocol), the United States, Japan, Australia, China, and a number of developing countries attended the Bali (Indonesia) conference. The members of the UNFCCC agreed that they would engage in a process of producing a comprehensive global climate agreement within the next two years. These countries have ample time to discuss proposals with different internal and external constituencies. Additionally, the countries are discussing various proposals for a future GHG emissions reduction regime, which may have a longer time horizon than the current five-year horizon. The key demand of the United States and other non-Kyoto Protocol countries is to bring emerging economies (China, Brazil, India, etc.) into the future carbon emissions reduction regime. At the same time, the emerging economies are eager to see the developed countries adopt aggressive carbon emissions reduction targets, provide technological assistance, and make capital available for modernizing developing country's carbon-intense industrial sectors. It is believed that the framework of a post-Kyoto Protocol regime may be agreed upon at the next UNFCC Conference of Parties (COP) meeting, in December 2009, to be held in Copenhagen, Denmark.

The economic picture is quite murky after 2012, when the mandatory requirement will end. A long-term policy will eliminate uncertainty and encourage corporations to invest in environmentally friendly technologies. Corporations need a longer-term horizon to plan their future investment activities; thus, the post-Kyoto2012 agreement should have a much longer time horizon than the current five-year arrangement. Most energy projects have a thirty- to fifty-year life span; thus, corporations need a long-term, consistent carbon regime.

Success of any future global climate control regime is linked to inclusion of China, Japan, India, and the United States in the post-Kyoto2012 framework, as well as some level of involvement by other emerging countries in future GHG emissions reduction targets. Furthermore, U.S. regional cap-and-trade agreements and any federal cap-and-trade system will also redefine the global climate control regime. Global corporations and financial markets must watch the level of targets and the timelines of future carbon regimes, domestic politics of large polluting countries, and the degree of coordination among separately managed global carbon regimes.

The level of targets and the timelines of the United States and other major polluting countries' carbon regimes will determine the supply and demand of future carbon products. A significant factor in the efficiency of the post-Kyoto2012 regime depends on harmonization of carbon reduction and verification standards across the countries and various regional carbon initiatives. In the present global carbon regime, the standards and their enforcement significantly vary among the CDM, JI, and different voluntary markets. In an ideal world, globally acceptable carbon emissions measurements and verifications standards will help the carbon market grow.

Australia, Canada, Japan, Mexico, South Korea, and New Zealand are also contemplating regional or federal cap-and-trade systems, but they are waiting for political developments in the United States and the outcome of the Copenhagen climate conference scheduled at the end of 2009. In September 2007, New Zealand set up a program to reduce all six GHGs for selected industries. Other industries can be gradually inducted into the system between 2008 and 2013. Australia's emissions trading program is under discussion and should be ready for legislation in 2009, so the possible date of implementation may be in 2010.

The trading program must weigh Australia's economic interests and the trading schemes of neighboring countries (Carbon Finance 2008a).

Japan has encouraged industries to voluntarily reduce emissions, rather than imposing GHG emissions caps. Japan is also planning an EU-style cap-and-trade mechanism, announced by the prime minister in June 2008, which could set the stage for a promising carbon market in the country. The city of Tokyo is also considering a cap-and-trade system to reduce GHG emissions (Chhabara 2008). Japanese industry associations have opposed the cap-and-trade plans, saying that tougher emissions targets will prompt businesses to shift production to developing countries. The Tokyo Stock Exchange is planning a carbon trading platform. Currently, Japanese corporations buy CERs directly from project developers in developing countries or through brokers, but a local exchange could reduce transaction costs. Details are still unclear about other countries' emissions trading systems.

Linkage of Carbon Markets

Linkage between different global mandatory carbon regimes, voluntary markets, and among U.S. carbon regimes is an important piece of the puzzle in the post-Kyoto2012 global carbon regime. By definition, linkage describes interchangeability among tradable permit systems (EU ETS, CDM, RGGI, and the proposed cap-and-trade regimes of Australia, Canada, Japan, New Zealand, Norway, Switzerland, and the United States) that allow emissions reduction efforts to be redistributed across global carbon regimes. Linkage among regimes helps corporations to trade their deficit or surplus carbon credits to meet their targets at a lower cost. This linkage among carbon regimes can be direct (unilateral or bilateral recognition of others' carbon products) or indirect (any two regimes that are linked to a third regime). An example of indirect linkage is the interchangeability of any two U.S. regional programs' carbon credits with CERs generated from CDM projects in developing countries. In this arrangement, the CER generatingCDM projects will establish an indirect link between these two U.S. regional programs, and, perhaps, carbon credits between these two U.S. regional programs may each be traded. Linkage is going to be the most critical dimension of any post-Kyoto2012 global carbon regime. For effective linkage framework, there have to be foolproof emissions monitoring systems, reporting and enforcement mechanisms, and harmonized standards across the carbon regimes.

There are numerous potential problems associated with the linkage mechanism. The linked carbon regimes will lose their freedom to modify their own targets and timelines, as they have to take all partners into confidence before making any change. Linkage will depend on the domestic policies of countries, and a successful linkage arrangement will reduce transaction costs of global climate exchanges and OTC markets. However, linkage will improve efficiency and reduce the cost of compliance in the developed world, where marginal cost of compliance is much higher than in the developing world. In a well-designed linkage system, offsetting and trading will be mutually beneficial to both developed and developing countries.

Carbon Neutrality

An encouraging trend of the global carbon market is the decision of several countries and corporations to become carbon neutral; the carbon neutrality trend will create demand for carbon credits and contribute to growth of the market. At the United Nations Climate Conference in February 2008, the United Nations Environment Program launched a new online network of countries engaged in the carbon-neutral endeavor. At the 154-nation meeting, Monaco, the host country, became the fifth to commit to carbon neutrality, joining Norway, New Zealand, Iceland, and Costa Rica. Carbon neutrality means that the countries will take steps either to drastically reduce their GHG emissions or the countries will offset their remaining emissions by purchasing credits in the global carbon market. Each country has proposed a timeline suitable to its economic condition, and the decision of these countries will eventually generate volume in the global carbon market. Norway plans to be carbon neutral by 2030. It is widely believed that Costa Rica, however, will emerge as the first carbon-neutral country, even though the country faces a host of economic problems, from illegal logging to overdevelopment fueled by tourism. On the positive side, Costa Rica's power generation sector is already nearly carbon neutral, and its small size may assist in better management of its GHG emissions (Lovgren 2008).

Another noteworthy development is the United Arab Emirates' decision to start building a carbon neutral city on February 10, 2008 (Dawn 2008). The new city's name is Masdar City, and it should be completed by 2015 with an estimated cost of $22 billion. The city will house fifty thousand people and will run entirely on renewable energy; residents will use electric-powered vehicles to move around. Many cities and localities, such as San Francisco, Toronto, Tokyo,

Copenhagen, and London have adopted aggressive carbon-cutting measures. A number of other cities have also announced their carbon neutrality goals; however, there is currently no comprehensive list of the cities desiring to become carbon neutral.

HSBC is the first global bank to declare a carbon neutrality goal in 2005. The bank has set up a 5 percent carbon emissions reduction target under three broad strategies. First, HSBC will manage and reduce its direct carbon emissions by implementing energy-saving plans. Second, the bank will increase the supply of renewable energy whenever possible. Third, the bank has started offsetting its remaining carbon emissions by purchasing carbon credits from green projects around the world. Similarly, a number of organizations, such as the World Bank, the World Resource Institute, and the Green Exchange, have adopted a policy of carbon neutrality. Global corporations are also announcing targets to reduce their carbon emissions, and there is a competition to be the first carbon neutral corporation within certain industries. In the technology industry, Dell, Inc. became the first carbon-neutral corporation in August 2008 (Dell 2008).

THE CARBON CAP-AND-TRADE SYSTEM IN THE UNITED STATES

A cap-and-trade system imposes a limit on how manyGHGs a corporation can emit. The corporation has to obtain a GHG emission permit, either through government auction or an award. The energy-efficient corporations can reduce their GHG emissions below the allowed limit of emissions, so they can trade (sell) their emission reductions to the corporations not able to reduce their emissions to the allowed level of GHG emissions. The cap-and-trade system serves as a "carrot" to the efficient corporations and a "stick" to the corporation not able to significantly reduce their GHG emissions.GHG emissions auction brings revenue to the governmental agencies as well.

The U.S. cap-and-trade (CAT) system is best known by the U.S. Senate's failed Lieberman-Warner legislation. The last version of the Lieberman-Warner legislation was called the America's Climate Security Act of 2007 (S-3036). The legislation collapsed on the senate floor on June 6, 2008, despite the fact that both presidential candidates in 2008 election supported this legislation. This was the first ever federal legislation proposal mandating carbon emissions reduction targets for key U.S. industries. The legislation was complex and put to vote prematurely (Fine and Bluestein 2008). It is widely believed that similar legislation will be discussed in 2009 by the new U.S. administration and Congress. The targets, timeline, sectors covered, targets allocation mechanism (auctions or free allocation), and percentage of domestic and international offsets are all going to be major components of any future U.S. CAT system.

The Lieberman-Warner legislation was introduced in October 2007; the Boxer's amendment to the Lieberman-Warner legislation was released on May 23, 2008, covering all six GHGs and offering an expected start date of 2012. Three major sectors included in this legislation were electricity production, transportation, and industry. The baseline was set up at the 2005 level of CO_{2e} emissions, and all GHGs were included in the proposed legislation. The suggested emissions cap in 2012 was between 5.2 and 5.7 billion tons of CO_{2e}, which wouldhave taken emissions back to the 2005 level. The proposal had two phases: 2012 through 2020 and 2021 through 2050. The ultimate target was to reduce GHG emissions to 70 percent below the 2005 baseline by the year 2050. The proposal had combined allocation and auction of the GHG reduction targets. In the early stage, allocation would comprise about 70 percent (decreasing annually), and auctions would comprise about 30 percent (increasing annually). All GHG reduction targets were to be auctioned off starting from 2031. The funds raised by auction were supposed to promote efficiency, new low-carbon technologies, and renewable energy.

One of the most debatable issues in any future CAT system in the United States is the offset flexibility. The offset provision allows corporations to trade low-cost carbon credits from various sources to high-cost reduction of GHG emissions from their current operation in developed countries. In the failed legislation, there was a cap of 15 percent trade with foreign credits (10 percent from international forestry credits and 5 percent from industrial projects), and 15 percent were allowed to be offset by domestic trade; this suggests that the included sectors would have had to reduce 70 percent of the assigned targets by employing carbon reduction processes, which could have been more expensive than the 30 percent offsets.

In any future U.S. CAT mechanism, the treatment of regional climate agreements and voluntary markets' credits will also be a tricky matter. One way to mitigate problems would be to create "early action" emissions allowance, which might partially reward the carbon credits generated before the enactment of the future federal legislation. At the same time, the regional climate regime standards may be difficult to harmonize with the future federal CAT regime because the regional standards may be more stringent standards than any federal mandate;

that is why it may be difficult to offset carbon credits under different regimes in the United States.

The U.S. House and Senate debated about ten pieces of environment-related legislationsintroduced in 2007, but the future of these legislations is unclear at this time. The U.S. House of Representatives started discussion on new GHG reduction legislation in early October 2008, and as of this writing, no progress has been reported on this piece of legislation (Point Carbon 2008). The new Congress will discuss any future U.S. CAT system in 2009 and may combine some of the aforementioned proposals. The U.S. politicians seem more eager to tackle U.S. and global environmental problems in 2009 than any time in the past. The future U.S. CAT system may include subsidies for several sectors, including research and development (R&D) activities, and the CAT system may coincide with the post-Kyoto2012 global carbon regime, witha starting date of January 1, 2013.The American Recovery and Reinvestment Act (signed by President Obama on February 17, 2009)has included funding for renewable energy, energy efficiency and other low-carbon industries. The American Recovery and Reinvestment Actis a positive step towards formulating a federal carbon cap-and-trade system.

In addition, there is another major proposal endorsed by the U.S. Climate Action Partnership. This group comprised large U.S. corporations with $1.7 trillion revenue in 2007. The proposal is called the U.S. Climate Action Program (USCAP), and it suggests a 60 to 80 percent carbon reduction by 2050. The group advocates technology solutions and flexibility along with carbon emissions-reduction targets.

Regional Carbon Initiatives

The actual nature of the U.S. federal cap-and-trade (CAT) system is still debatable, but there are several regional initiatives that are much more specific in nature.

1. **Regional Greenhouse Gas Initiative**
   The first initiative is the Regional Greenhouse Gas Initiative (RGGI), a cap-and-trade program. Ten states and more than one hundred mayors of U.S. cities have signed this agreement, which is to be implemented in 2009. The program focuses only on power plants larger than 25 MW, and involves an initial compliance period from 2009 to 2018, cutting

emissions by 10 percent. The draft agreement was signed in December 2005 by ten northeastern and midAtlantic states. The participating states maintain their right to include additional carbon-polluting sectors after reducing GHG emissions from large power plants (RGGI 2008). The overall compliance by 2014 is targeted at 188 million short tons of CO_{2e}, which are about 8 million short tons of GHG above the 2000–2002-emissions level (baseline). The cap is kept high enough to accommodate future growth, one of the concerns of business groups, but critics argue that the cap is not aggressive enough. RGGI-participating states have already started auctioning their GHG allowance; they've been doing so since the end of September 2008 (RGGI 2008). RGGI futures and options contracts were launched on the Chicago Climate Futures Exchange's (CCFE) launch on August 15, 2008 (Point Carbon Newsletter August 15, 2008). The ten participating states are Connecticut, Delaware, Maine, Maryland, Massachusetts, New Hampshire, New Jersey, New York, Rhode Island, and Vermont.

2. **Western Climate Initiative**

   The second regional initiative in the United States is the Western Climate Initiative (WCI), which was launched in February 2007. As of January 2009, the participating members of the WCI are Arizona, California, Montana, New Mexico, Oregon, Utah, Washington, and four Canadian provinces (British Columbia, Manitoba, Ontario, and Quebec). Canadian province of Saskatchewan, six U.S. states, and several Mexican provinces have maintained observer status with the WCI.

   WCI is a broad cap-and-trade system covering a wide range of industrial methodologies, transportation, and home consumption and targeting ninety emissions from the partner states and provinces. The goal of the WCI is to reduce GHG emissions of the region15 percent from the 2005 level by a deadline in 2020. The GHG emissions cap will take effect from 2012, so the first compliance phase is 2012 to 2020. This target was announced in August 2007, but the details need to be worked out before compliance is mandated (WCI 2008). The process of GHG allowance auctioning or allocation has not started yet.

3. **Midwestern Greenhouse Gas Accord**

   The third initiative is the Midwestern Greenhouse Gas Accord (MGGA) signed by ninemidwesternstates (Illinois, Indiana, Iowa, Kansas, Ohio,

Michigan, Minnesota, South Dakota, and Wisconsin) and the Canadian province of Manitoba. The initial accord was signed in by six U.S. states and Manitoba in November 2007, with a desire to implement a regional multi-sector cap-and-trade program in the region by July 15, 2010 (Doyle 2008 and Midwest Governors Association 2007). The target is to reduce GHG emissions by 60 to 80 percent of the 2007 level, and the deadline is 2050. The MGGA plans to include several industries, and it is expected that other midwesternstates will join as well.

These three U.S. regional initiatives account for about one-third of overall U.S. GHG emissions; that is why the three regional initiatives alone cannot solve the country's environmental problem.However, these three regions are leading the GHG management struggle in the United States. These regional initiatives may provide some direction to the U.S. federal government in designing future cap-and-trade systems. The share of CO_{2e} emissions only from the states participating in these three regional initiatives is summarized in Table 9.1. RGGI accounts for only 10 percent of the total U.S. emissions, and emissions from the WCI and MGGA member states account for 13 percent and 14 percent.

Table 9.1

Total Emissions and Percent of Total U.S. Emissions from Regional Initiatives

Regional Initiative	**Participating U.S. States' Total Emissions** (million metric tons of CO_{2e})	**Percent of Total U.S. Emissions**
Regional Greenhouse Gas Initiative	695	10%
Western Climate Initiative	871	13%
Midwestern Greenhouse Gas Accord (original six states)	932	14%

Source: (Damassa 2007)

Module 10

Fifty Steps to Reduce Your Carbon Footprints

THIS section provides a number of simple lifestyle changes and steps that can help reduce carbon footprints, save money, and possibly improve quality of life. The list is a suggestion, and readers can choose the steps that may be suitable to their ways of life. This is not an exhaustive list, and readers are encouraged to learn more about reducing carbon footprints. A list of selected Websites, where readers can find additional details, is available in Table 10.1.

While on the Road

1. You can avoid aggressive driving and reduce carbon footprints as well as improvingyour safety (and that of those around you). Try to maintain the legal speed as long as traffic allows. Less aggressive driving (rapid acceleration and braking) improves gas mileage.
2. It is a good idea to focus on your vehicles' maintenance and regularly check tire pressure. According to the U.S. Department of Transportation's analysis, properly inflated tires and clean air filters significantly reduce gas consumption and, as a consequence, reduce carbon footprints.
3. An idling vehicle emits many-fold more pollutants into the air than the emissions of a moving vehicle, and idling lowers mileage per gallon. Thus, you should avoid idling your vehicle as much as possible while you are stopped at a stoplight or waiting for someone.
4. Public transportation can be another alternative during heavy-traffic times. If public transportation is not suitable to your lifestyle, it might be a good idea to rearrange work hours to avoid driving during heavy-traffic times, since driving in heavytraffic consumes nearly twice as much fuel

than normal-speed driving. By using public transportation and avoiding heavy-traffic times, you can reduce your carbon footprint.
5. Carpooling may not be suitable for everybody's lifestyle, but planning ahead and combining multiple trips can save you up to one thousand dollars annually in fuel consumption and vehicles maintenance. Carpooling and well-planned trips will reduce carbon footprints.
6. Keeping your vehicles a little less warm in winter and a little less cool in summer is another lifestyle change that you might consider implementing. This habit will improve your gas mileage. Keeping windows open (as an alternative to using a vehicle's cooling system) during low-speed driving when the outside temperature is between 60° F and 75° F will likewise reduce your carbon footprint.
7. You do not have to purchase expensive, high-octane gasoline if it is not recommended in your vehicle-owner's manual. The few cents cheaper gasoline may work as well as the high-octane gasoline and lower carbon footprints.
8. More and more auto corporations are rolling out hybrid models every year. It is highly recommended to buy hybrid vehicles; the fuel savings will balance out the slightly higher purchase price. Leading auto corporations such as Toyota, General Motors, and Chrysler have announced the roll out of cost-effective electric cars in the near future, so keep an open mind about electric cars or other fuel-efficient options. The federal government offers a tax break for those who purchase these low-carbon-footprint vehicles. Above all, you will get satisfaction from supporting the new energy-efficient technologies.

While Indoors

9. You can keep your home thermostat two degrees Fahrenheit lower in winter and two degrees Fahrenheit higher in summer than what you are used to. This small change will not drastically influence your lifestyle, but it will cut down both carbon emissions and utility bills.
10. You can purchase energy-efficient appliances and unplug them when they are not in use. Additionally, you should run full loads in dishwashers and washing machines. You can also cut down your carbon footprint by setting cool or warm water settings on your dishwasher and washer.
11. Whenever possible, you should purchase or replace appliances with

energy-efficient appliances. Ask an appliance salesperson about the Energy Star appliances.

12. You can install low-flow plumbing in bathrooms and reduce water consumption. Water purification by city governments is an energy-intense operation. Also, in the future, water is going to be as valuable and scarce as petroleum. Water allocation is already a point of contention among many countries. You can start saving money and reduce your carbon footprint by reducing water consumption.
13. You can reduce your carbon footprint and your water bill by taking shorter showers and turning off faucets when you don't need water. Every two minutes in the shower uses about ten gallons of water. Cutting down the frequency of bathing in the bathtub may be a good idea because bathing consumes much more water than a shorter shower.
14. These days, most foods are industrially processed or packed and then transported thousands of miles. All these activities consume fossil fuel and increase consumer's carbon footprints. If we all pledge not to waste cooked or uncooked food, not only will consumers save money and reduce carbon footprints, we will make a dent in global food prices.
15. Use microwaves as often as possible because microwaves consume less power than electric or gas ovens. Also, keeping a microwave clean will further improve its efficiency and consume much less power.
16. Utility industries are the biggest polluters in the world. Reducing electric power use will not only save you money, it will help utility corporations to consume less fossil fuel. Keep light bulbs clean and turn off lights when you are leaving your house or office.
17. It is a good idea to gradually change your light bulbs to compact fluorescent bulbs. These bulbs are more expensive, but they consume about one-fourth of the electricity of incandescent bulbs, and they last much longer.
18. Holiday celebration habits may need some modification. Using LED-type holiday lights will reduce your electricity bill and reduce carbon footprints.
19. Most newlybuilt houses are well insulated and energyefficient; however, older houses may not be as energyefficient as we want. Make sure that any minor openings in windows and doors are sealed, particularly during cold winter or hot summer months. A simple test to find small

air leaks is to move a candle next to window and door frames. If the candle flickers, try to locate the opening and seal it professionally. The American Recovery and Reinvestment Act of February 17, 2009 have allocated billions of dollars for Americans to better insulate their homes and offices.

20. Investing in improving home insulation and energy efficiency techniques may balance out cost in the long run and reduce carbon footprints.
21. An easy lifestyle change is to start covering windows with drapes and curtains to help reduce heating and cooling costs and to protect yourselves from sunrays.
22. Keep all doors, windows, and ventilators closed when cooling and heating systems are on.
23. In cold locations, servicing your furnace annually is highly recommended. It may be a good idea to replace older furnaces with a more efficient furnace such as a condensing model.
24. If your heating and cooling systems allow, only heat or cool the spaces in use. Placing the cooling or heating system's thermostat in a location that is neither hotter nor cooler than the rest of the building will better regulate temperature and reduce carbon footprints.
25. Another easy way to reduce utility bills is to keep your refrigerators and freezers clean, as well as keeping these appliances several inches away from the wall for air circulation and efficient operation.
26. Cooking several meals at one time can reduce your utility bill and food waste, while utilizing time for other activities.
27. Consume a minimum of office supplies, even if they are free to you. This will save your employer money and contribute less waste to landfills.
28. Using both sides of paper in your copying and printing tasks will save youmoney, and fewer trees will be processed. You can complete your copying and printing tasks to your satisfaction while reducing carbon footprints.
29. There are ways to put a hold on the junk mail that you receive at home or work. Mass advertisers waste tons of paper. Refer to Table 10.1 for Websites where you can initiate the process. You can also avoid or recycle telephone books.
30 Net metering laws (consumers' right to sell excess electricity to utility companies) in many Americanstates and falling prices for wind and

solar energy equipment can be viewed as an invitation to many homeowners to become energy producers. An investment in home-installed wind turbines, solar panels, and solar water heaters should be paid off by reduced utility bills and possible profits from selling surplus power to local utility corporations as allowed by net metering laws.

While Outdoors or Shopping

31. An easy way to save money and reduce your carbon footprint is to focus on your backyards and patios. About one-third of all fertilizers and summer household water in the United States is consumed by maintaining backyards and patios. You can better educate yourselves on responsible gardening. One of the easy solutions is to spread out cut grass clippings and keep grass at least two inches long. This action alone will protect your backyards and patios from heat, resulting in reduced need for water and fertilizer.
32. Plant trees around your house. Trees not only absorb CO_2 from the air; they also provide valuable shade if planted at the right location, which can partially cool off your buildings and reduce cooling costs.
33. Shopperscan take their reusable bags to the store and avoid using the store's shopping bags. Obviously, consumers have different needs, but carbon-conscious shoppers can prioritize their shopping bag use as follows: use of no bags, use of as few bags as needed, use of biodegradable paper bags, or use of regular plastic bags. Also, you can reuse shopping bags for your other needs.
34. If you can foresee your needs for a couple of weeks ahead, you should consider buying in bulk or shopping at wholesale stores. Large and bulk packaging waste less paper and packing material per unit, and they are considerably cheaper. The savings can be as high as one-half of the small (one-serving) packaging.
35. Buy seasonal fruits and vegetables. They are full of nutrients, lower in price, free of industrial processing, and do not require excessive transportation costs, thus reducing carbon footprints.
36. Buy locally produced products whenever possible. This will lower transportation-related carbon emissions.
37. Resist the temptation to purchase new clothes and consumer goods every few months. The industrial sector is a major greenhouse gas (GHG)

emitter, and the industry consumes valuable water resources as well. You can save money by purchasing fewer clothes and consumer goods (and still maintain your lifestyle) and recycling old items at a local charity or shelter.

38. Cellular phone providers around the world are rolling out new phone models each year and are aggressively marketing new models to consumers. Resist the temptation to continually purchase new phones, and save money while reducing your carbon footprint. You can safely skip the next new model without missing out on useful features.

39. In light of rising fuel costs, business and personal commuting should be kept at a minimum; minimize commuting and combine trips whenever possible. Low-carbon commuting will save money and reduce carbon footprints. Some of the airlines are offering carbon neutral travel options at a higher cost. The airlines are pledging to purchase carbon credits generated from green projects to offset your share of GHG emissions.

40. Consider renting out sporting equipment if you are not an avid sportspersons. Most sports centers rent out all kinds of sport equipment. Sporting equipment is expensive, and manufacturing these products is carbon intense. Renting is a cheaper and low-carbon solution for most consumers.

41. If possible, consumers can reduce or replace over-the-counter medicine with herbal and homeopathic medicines. Homeopathic and herbal medicines are less expensive, and their production is less harmful for the environment. Consumers can educate themselves about holistic medicines and treatments. A list of informativeWebsites is provided in Table 10.1.

42. Citizens should rethink their gift-giving habits and social-interaction modes. You can save money and lower your carbon footprint at the same time by not purchasing greetings cards and not purchasing gifts with expensive, bulky packaging. Purchasing quality gifts online may be another alternative that saves money and reducesyour carbon footprint.

43. If there is no compelling need, purchase notebook rather than desktop computers. Notebooks require less material to build and save twenty to forty dollars in electricity per year, depending on the usage and location of the residence.

44. Whenever possible, give treadmills and gym equipment a break, and go

for a walk or jog outside. Outdoor activities have positive psychological benefits and save electricity (thus lowering carbon footprints).

45. If all of you try to take reusable water bottles and clean or filtered water from home with you, not only will you save hundreds of dollars each year, but you will also save pounds of plastic. Take water to work, social events, and workout activities. The bottled water industry is a major polluter.

46. Online banking and electronic payment options are recommended. You can save over one hundred dollars on postage each year, avoid late fees, and save paper. Online banking also saves valuable time and transportation costs and, obviously, reduces your carbon footprint.

Strategic Steps

47. Online education has a potential to reduce students' carbon footprints. Online education will save valuable driving time, commuting costs, and textbook costs for most traditional and adult learners. However, online education should be a personal choice because it may not be suitable to everyone's learning style. Also, developing an online education curriculum is a strategic challenge to educational institutions, and online education may not be a viable option in all academic disciplines.

48. It is understood that investors are interested in higher returns. Whenever possible, invest in green and socially responsible corporations. These corporations have low-carbon compliance risks, and these corporations are quite often run by responsible management. Green and socially responsible corporations have less risk of being sued by the government or consumers. Investing in green corporations and mutual funds whenever possible can help investors earn consistent returns and help improve the environment. Module 5 has outlined some of these green investment choices.

49. You can be proactive and work with local and federal politicians to come up with innovative solutions to reduce your communities' carbon footprints.

50. You can be proactive and work with your employers to come up with innovative solutions to reduce your employer's carbon footprint by suggesting flexible work hours and other flexibilities.

Table 10.1

Selected Websites onReducing Carbon Footprints

100 Ways to Conserve
http://www.wateruseitwisely.com/100ways/index.shtml
APS Green Choice
http://www.aps.com/main/green/choice/default.html
Consumer Energy Center
http://consumerenergycenter.org
ConsumerReports.Org
http://www.consumerreports.org/cro/money/
Demystify Homeopathy
http://www.demystify.com/index.htm
Eco-Cycle
http://www.ecocycle.org/junkmail/index.cfm
Energy Kids Page
http://www.eia.doe.gov/kids/
Energy Star
http://www.energystar.gov/
EnergyTrust of Oregon, Inc.
http://www.energytrust.org/
Green Dimes
http://www.greendimes.com/
Holistic Healing Web Page
http://www.holisticmed.com/index.shtml
Index of University of California, Berkeley Research
http://www.ocf.berkeley.edu/~sgb/pledge/pledgeinfo/
National Coalition for the Homeless
http://www.nationalhomeless.org/publications/facts.html
Natural Resources Defense Council
http://www.nrdc.org/water/
Saving Energy
http://www.energyquest.ca.gov/saving_energy/change_light.html
Sierra Club
http://sierraclub.org
Stopjunkmail.com

http://www.stopjunkmail.com

Summary of the Financial Benefits of Energy Star Labeled Office Buildings
http://www.csemag.com/contents/pdf/070523JeanESsummary052307.pdf

The International Ecotourism Society
http://www.ecotourism.org/webmodules/webarticlesnet/templates/eco_template.aspx?a=12&z=25

U.S. Census Bureau (Quick Facts)
http://quickfacts.census.gov/qfd/states/00000.html

United Nations Environment Programme
http://www.unep.org/geo2000/ov-e/index.htm

What is Recycling?
http://www.ilacsd.org/recycle/recycle.php

Appendix 1

Selected Acronyms

AA	Assigned Amount
AAU	Assigned Amount Unit
ACX	Asia Carbon Exchange
AIE	Accredited Independent Entity

BAU	Business As Usual
BM F	Bolsa de Mercadorias & Futuros, Brazil

CAT	Cap-and-Trade
CCAR	California Climate Action Registry
CCE	Canadian Climate Exchange, Inc.
CCFE	Chicago Climate Futures Exchange
CCS	Carbon Dioxide Capture and Storage
CCX	Chicago Climate Exchange
CDM	Clean Development Mechanism
CDM EB	Clean Development Mechanism Executive Board
CDP	Carbon Disclosure Project
CERs	Certified Emission Reductions

CFU	Carbon Finance Unit (The World Bank Group)
CIF	Climate Investment Funds
CO2e	Carbon Dioxide Equivalent
COP	Conference of the Parties
CPF	Carbon Partnership Facility

DNA	Designated National Authority
DOE	Designated Operational Entity

EEX	European Climate Exchange
EIT	Economies In Transition
EPA	Environmental Protection Agency
ER	Emission Reductions
ERPA	Emission Reduction Purchase Agreement
ERPA	Emission Reduction Purchase Agreement
ERU	Emission Reduction Unit
E-t-C	Emission to Cap
EU ETS	European Union Emissions Trading Scheme
EUA	European Union Allowance

FCPF	Forest Carbon Partnership Facility

G5	Group of five countries
G77	Group of seventy-seven countries
GHGP	Greenhouse Gas Protocol

Selected Acronyms

GHGs	Greenhouse Gases
GIS	Green Investment Scheme
GWP	Global Warming Potential

IBRD	International Bank for Reconstruction and Development
IDA	International Development Agency
IEA	International Energy Association
IER	International Emission Trading
IFC	International Finance Corporation
IIGCC	Institutional Investors Group on Climate Change
IPCC	Intergovernmental Panel on Climate Change
IPO	Initial Public Offering
IRR	Internal Rate of Return
ISO	International Organization of Standardization
ITL	International Transaction Log

JI	Joint Implementation
JISC	Joint Implementation Supervisory Committee

LCERs	Long Term Emission Reductions
LOA	Letter of Approval
LOE	Letter of Endorsement
LOI	Letter of Intent
LULUCF	Land Use, Land Use Change and Forestry

MAC	Marginal Abatement Cost
MCEx	Montreal Climate Exchange
MCX	Multi Commodity Exchange-India
MDB	Multilateral Development Bank
MGGA	Midwestern Greenhouse Gas Accord
MOP	Meeting of Parties
MOU	Memorandum of Understanding
MP	Monitoring Plan

NAP	National Allocation Plan
NGO	Non Governmental Organization
NSF	National Science Foundation
NYMEX	New York Mercantile Exchange

OE	Operational Entity
OTC	Over the Counter

PCN	Project Concept Note
PDD	Project Design Document
PIN	Project Idea Note
PPA	Power Purchase Agreement

REC	Renewable Energy Credit
RGGI	Regional Greenhouse Gas Initiative

Selected Acronyms

RMU	Removal Unit

S&P	Standard & Poor

UNFCCC	United Nation Framework Convention on Climate Change
USCAP	United State Climate Action Program

VC	Venture Capital
VCU	Voluntary Carbon Unit
VERs	Verified Emission Reduction
VMS	Voluntary Market Standards

WBCSD	World Business Council on Sustainable Development
WCI	Western Climate Initiative
WRI	World Resource Institute

Selected References

Asplund, R. W. *Profiting from Clean Energy: A Complete Guide to Trading Green in Solar, Wind, Ethanol, Fuel Cell, Carbon Credit Industries, and More.* Hoboken, New Jersey: Willey & Sons, 2008.

AWEA. American Wind Energy Association, www.awea.org. 2008 (accessed October 11, 2008).

Ball, Jeffrey. "Six Products, Six Carbon Footprints." *The Wall Street Journal,* October 6, 2008 (2008a): R1, R3, R4.

Ball, Jeffrey. "Pollution Credits Let Dumps Double Dip." *The Wall Street Journal,* October 20, 2008 (2008b): A1, A14.

BBC News. U.S. Agrees to Climate Deal with Asia. July 28, 2005, http://news.bbc.co.uk/1/hi/sci/tech/4723305.stm.

Brahic, Catherine. "China's Emissions May Surpass the US in 2007." *New Scientists*, April 2007, http://environment.newscientist.com/article/dn11707-chinas-emissions-to-surpass-the-us-within-months.html, (accessed August 20, 2008).

Carbon Finance Unit. www.carbonfinance.org. 2008 (accessed August 17, 2008).

Carbon Finance. "Australia Specifies ETS Design, Timetable." *Carbon Finance* February 2008 (2008a): 13.

Carbon Finance. "First Indian Carbon fund Closes." *Carbon Finance* March 2008 (2008b): 12.

Carbon Finance. "NYMEX Launch First Suite of Carbon Contracts." *Carbon Finance* March 2008 (2008b): 11.

Carbon Footprint. 2008. http://www.carbonfootprint.com, October 11, 2008.

Carbon Rating Agency. www.carbonratingsagency.com, 2008 (accessed July 29, 2008).

Carbon Trust. www.carbontrust.com/EN/Home.aspx, 2008 (accessed October 5, 2008).

CCE. Canadian Climate Exchange.www.canadianclimateexchange.com/, 2008 (accessed July 29, 2008).

CCX. Chicago Climate Exchange, www.chicagoclimatex.com, 2008 (accessed January 11, 2009).

CDM. Clean Development Mechanism, http://cdm.unfccc.int/index.html, August 11, 2008.

Chernova, Yuliya. "Consumers as Producers, "*The Wall Street Journal*, November 17, 2008: R4.

Chhabara, Raesh. "Asia Switches on the Carbon Market."*ClimateChangeCorp: Climate News for Businesses* July 28, 2008, http://www.climatechangecorp.com/content.asp?contentid=5518, (accessed November 17, 2008).

CleantechNetwork , www.cleantechnetwork.com(accessed January 17, 2009)

Cundy, Christopher. "Options on Exchanges."*Environmental Finance,* March 2008:16–17.

Damassa, Thomas. "The Midwest Greenhouse Gas Initiative by Numbers, "*World Resource Institute*, November 21, 2007.

Dawn. "U.S. Agrees to Join Talks on Climate Change: Kyoto Pact Extended Beyond 2012." *Dawn*, December 11, 2005.

Dawn. "UAE Starts Work on Zero-Carbon City."*Dawn*, February 11, 2008.

Dell. www.dell.com, June 29, 2008.

DJSI. The Dow Jones Sustainability Index, www.sustainability-index.com, 2008 (accessed July 17, 2008).

Doyle, Jim. "A New Direction, "*Trading Carbon* 2:2 (2008): 40–42.

ECX. European Climate Exchange, 2008.www.europeanclimateexchange.com.

Eilperin, Juliet. "Climate Plan Splits U.S. and Europe."*Washington Post*, July 2, 2005.

Energy Information Administration."International Carbon Dioxide Emissions and Carbon Intensity, " Energy Information Administration: Official Energy Statistics from the U.S. Government, 2008, www.eia.doe.gov/emeu/international/carbondioxide.html, (accessed July 17, 2008).

Environmental Finance. "CCX, Montreal Team Up for Climate

Exchange." *Environmental Finance*, 2008, www.environmental-finance.com(accessed December 8, 2005).

Environmental Leader. "Voluntary Carbon Market Tripled in 2007, Hit $331 M." *Environmental Leader: The Executive's Daily Green Briefing*, http://www.environmentalleader.com, (accessed May 14 2008)

EPA. Environment Protection Agency, 2005, www.epa.gov/climatechange/emissions/downloads/08_CR.pdf.

Fine, Steven and Joel Bluestein."Life after Lieberman-Warner, "*Carbon Finance,* July 2008: 15.

GHGPI. The Greenhouse Gas Protocol Institute, www.ghgprotocol.org, (accessed August 11, 2008).

Gold Standard, www.cdmgoldstandard.org, (accessed August 20, 2008).

Green Exchange, www.greenfutures.com, (accessed August 11, 2008).

Green VC: News and resources on green venture capital, funding, and startups, www.greenvc.org/, July 29, 2008.

Greenhouse Gas Inventory Data, http://unfccc.int/ghg_data/items/3800.php, (accessed July 29, 2008).

Harvey, Fiona. "Carbon Trading Set to Dominate Commodities."*Financial Times,* June 25, 2008.

Hodge, Nick. "How to Use Green Venture Capital for Personal Profit."*Green Chip Stocks*, www.greenchipstocks.com/articles/green-venture-capital/245, June 10, 2008.

IEA. International Energy Agency, www.iea.org, September 7, 2008.

IIGCC. Institutional Investor's Group on Climate Change, www.iigcc.org, August 29, 2008.

Innovest, www.innovestgroup.com, July 17, 2008.

IPCC. "Climate Change 2007: Synthesis Report." Intergovernmental Panel on Climate Change, United Nations Environmental program and World Meteorological Organization, Geneva, 2007, www.wunderground.com/education/ipcc2007.asp (accessed January 17, 2009)

Jacquot, Jeremy E. "World Banks Plan Aims to Reduce Deforestation and Forest Degradation."*Treehugger*, http://www.treehugger.com/files/2007/12/world_bank_plan.php, (accessed October 6, 2008).

Khanna, Naresh. "Climate Exchange Breakthrough in Chicago, "*Indianprinterpublisher.com*, January 2003.

Kyoto Protocol. http://unfccc.int/kyoto_protocol/items/2830.php, December 31, 2008.

Lovgren, Stefan. "Costa Rica Aims to be First Carbon-Neutral Country."*National Geographic News* March 7, 2008.

McCarthy, Ryan J. "Venture Capitalists Flock to Green Technology."*Inc. com* March 28, 2006.

MCeX. Montréal Climate Exchange 2008, www.mx.ca/accueil_en.php?changeLang=yes&(accessed May 23, 2008).

Merrill, Amy and Jain Vivek. "Europe Leads way in New Era of Carbon Trading." *International Financial Law Review*24:9 (2005): 47–49.

Midwest Governors Association, http://www.midwesterngovernors.org/govenergynov.htm(accessed December 29, 2007).

Montgomery, Robert. "Managing Risk and Adding Value."*Environmental Finance* March 2008: 34–35.

Murray, Alan. "Next President Needs to Uncap Debate on Cost of Emissions Curbs."*The Wall Street Journal* March 17, 2008: A2.

New Carbon Finance.http://www.newcarbonfinance.com (accessed January 27, 2009)

Point Carbon. www.pointcarbon.com (accessed October 11, 2008).

Point Carbon Newsletter. 2008. www.pointcarbon.com (accessed August 15, 2008).

Power, Stephen. "EPA Says Carbon Caps Won't Harm Economy Much."*The Wall Street Journal*March 17, 2008: A2.

RGGI. Regional Greenhouse Gas Initiative, 2008, http://www.rggi.org/ (accessed August 2, 2008).

Sharman, Tom. "Lack of Progress at Poznan."*Guardian*, www.guardian.co.uk, December 12, 2008.

Simms, Andrew, Julian Oran and P. Kjell. *The Price of Power: Poverty, Climate Change, the Coming Energy Crisis and the Renewable Revolution,* Published by New Economic Foundation (www.neweconomics.org), London, June 2004.

SMH. "Sydney Declaration on Climate Change and Energy."*The Sydney Morning Herald*, September 9, 2007, http://www.smh.com.au/news/.

Taylor, Edward. "Start-Ups Race to Produce 'Green' Cars."*The Wall Street Journal* May 6, 2008: B1, B7.

The Green Power Network. U.S. Department of Energy, 2008 http://

erendev.nrel.gov/greenpower/markets/pricing.shtml?page=1(accessed July 29, 2008).

The World Bank.The World Bank Group, www.worldbank.org/climate-consult), (accessed August 17, 2008).

UNEP. United Nations Environment Program, http://www.unep.org/publications/ebooks/kick-the-habit/Default.aspx?bid=ID0E2PAI, August 29, 2008.

UNFCCC. Greenhouse Gas Emission Data for 1990–2003. *United Nations Framework Convention on Climate Change*, http://unfccc.int/ghg_data/ghg_data_unfccc/items/4146.php, December 20, 2005.

UNFCCC. 2008. *United Nations Framework Convention on Climate Change*, http://unfccc.int/2860.php, September 29, 2008.

VCS. Voluntary Carbon Standards.www.v-c-s.org (accessed January 19, 2009).

Victor, David and Joshua House. "A New Currency: Climate Change and Carbon Credits." *Harvard International Review* Summer 2004: 56–59.

WCI. Western Climate Initiative, 2008, www.westernclimateinitiative.org(accessed August 19, 2008).

WRI. World Resource Institute, 2008, www.wri.org(accessed September 17, 2008).

WSJ. "Global Warming: a Cloudy Outlook." *The Wall Street Journal*December 10, 2005.